ACS SYMPOSIUM SERIES **755**

Specialty Monomers and Polymers

Synthesis, Properties, and Applications

Kathleen O. Havelka, EDITOR
The Lubrizol Corporation

Charles L. McCormick, EDITOR
University of Southern Mississippi

American Chemical Society, Washington, DC

Chemistry Library

Library of Congress Cataloging-in-Publication Data

Specialty monomers and polymers: synthesis, properties, and applications / Kathleen O. Havelka, editor, Charles L McCormick, editor.

    p. cm — (ACS symposium series, ISSN 0097–6156; 755)

Includes bibliographical references and index.

ISBN 0–8412–3637–2

1. Polymers—Congresses. 2. Monomers—Congresses.

I. Havelka, Kathleen O., 1964– . II. McCormick, Charles L., 1946– . III. Series.

QD380 .S865   2000
547′.7—dc21
                                         99–58314

Foreword

THE ACS SYMPOSIUM SERIES was first published in 1974 to provide a mechanism for publishing symposia quickly in book form. The purpose of the series is to publish timely, comprehensive books developed from ACS sponsored symposia based on current scientific research. Occasionally, books are developed from symposia sponsored by other organizations when the topic is of keen interest to the chemistry audience.

Before agreeing to publish a book, the proposed table of contents is reviewed for appropriate and comprehensive coverage and for interest to the audience. Some papers may be excluded in order to better focus the book; others may be added to provide comprehensiveness. When appropriate, overview or introductory chapters are added. Drafts of chapters are peer-reviewed prior to final acceptance or rejection, and manuscripts are prepared in camera-ready format.

As a rule, only original research papers and original review papers are included in the volumes. Verbatim reproductions of previously published papers are not accepted.

ACS BOOKS DEPARTMENT

Contents

APPLICATIONS

Preface

This book is a compendium of refereed papers based on invited talks presented at the American Chemical Society (ACS) Symposium on Specialty Monomers and Their Polymers. Specialty monomers are inherently different from commodity monomers, and it is these differences that result in new and important polymer properties. The most versatile feature of specialty monomers and polymers is their built-in functionality. This functionality offers significant opportunity to customize the design of polymers to meet specific performance, physical, and chemical properties. These properties are needed for many of today's hi-tech applications and cannot be easily obtained with more common commodity monomers and polymers.

Specialty polymers have emerged as an important class of materials because they offer flexibility, both at the molecular and bulk levels, to optimize physical and chemical properties for medical and industrial applications. Because the structure of synthetic specialty polymers can be easily controlled and measured, they are frequently used to model the more complex natural polymers. Using well defined model compounds facilitates significant advancements in our understanding of fundamental properties. Increased understanding of materials issues will expedite the development of "smart polymers" that have unique, tunable, and responsive properties. Developing "smart polymers" requires a highly multidisciplinary approach, the progress of which depends on active collaboration from diverse groups, including chemists, biologists, physicians, physicists, material scientists, and engineers.

Considering that interest in specialty polymers and "smart polymers" is growing worldwide and considering its multidisciplinary nature, a need exists for a comprehensive volume that covers important current developments in this field. This book is intended to fill such a void by providing broad coverage of recent work by leading international scientists in their respective areas. The book is comprehensive in scope, covering and providing the state of the art in topics ranging from polymer synthesis, to materials, to applications. We intend that this book will be of great value to researchers entering the area as well as a valuable reference book for researchers of varied backgrounds.

We acknowledge the overwhelmingly positive response from the specialty polymers community. The breadth of participation in the symposium is due in part to the support from a number of organizations. In particular, we thank the ACS Division of Polymer Chemistry, Inc.; the Lubrizol Corporation; and Donors of the Corporation Associates and the Petroleum Research Fund, administered by the ACS for their financial support.

We are truly indebted to the numerous authors for their timely effort and to the referees for their critical evaluation of the manuscripts.

KATHLEEN O. HAVELKA
The Lubrizol Corporation
29400 Lakeland Boulevard
Wickliffe, OH 44092–2298

CHARLES L. MCCORMICK
Department of Polymer Science
University of Southern Mississippi
Hattiesburg, MS 39046–0076

OVERVIEW

Chapter 1

Specialty Polymers: Diverse Properties and Applications

Kathleen O. Havelka

The Lubrizol Corporation, Research and Development,
29400 Lakeland Boulevard, Wickliffe, OH 44092–2298

Many of the most accessible commodity polymers have been studied in detail and their applications have been developed extensively. Interest is now focused on the synthesis and study of specialty polymers to meet today's critical applications. Specialty polymers are primarily water-soluble polymers with functional groups that are pendent to or on the backbone; their global market value currently exceeds 9 billion dollars and continues to grow (1). They are used extensively for their unique solution properties in various fields of industry, agriculture, medicine, biotechnology, and electronics. Applications of specialty polymers are diverse, including: water treatment, paper processing, mineral sequestering, textile processing, personal care products, pharmaceuticals, drug delivery, petroleum production, enhanced oil recovery, coatings and inks additives, and sensors. Specialty polymers often provide multifunctional properties, i.e., a thickener for a paint may also act as a dispersant for the pigment in the paint. The level of multifunctional properties depends on the type of polymer, the amount used, and interactions with other chemistries in the formulation.

The diverse properties of specialty polymers are capturing the imagination of scientists and engineers worldwide because of their potential applications in many areas of present and future hi-tech and biological applications. Of particular interest is the ability to tailor a polymer to deliver a specified property in response to an external stimuli. This is highly desirable as it enables one to minimize both the amounts of material used and competitive reactions, which is critical in control-release functions, particularly in drug delivery applications. Effective delivery of drugs to a target cell or tissue largely diminishes adverse side effects and increases the pharmacological activity. It is the potential to tailor macromolecules to provide specified properties in an environmentally benign manner that motivates much of the current applied and fundamental research in the field of specialty polymers. These properties are critically needed for many of today's demanding applications.

Material Considerations of Specialty Polymers

There are numerous considerations in determining the appropriate polymer for a given application. In general, for a polymer to be commercially viable it must meet stringent environmental regulations, high performance standards, and be cost effective. The properties of specialty polymers depend on their basic chemical and structural properties, as summarized below.

- Molecular Weight – The molecular weight and molecular weight distribution strongly affect the solution properties of specialty polymers and their applications.
- Hydrophile to Lipophile Balance (HLB) -- the type and content of hydrophiles and lipophiles significantly impacts specialty polymers inter- and intra-molecular associations.
- Sequence distribution of monomers -- block, alternating, or random distribution impacts interaction of monomer groups within the polymer.
- Degree of Branching -- a branched polymer frequently has different properties than its linear analog, such as, lower tendency to crystallize, different solution properties and light scattering behavior.
- Hyperbranched materials and dendrimers -- offer the synthetic ability to chemically tailor the branches in a step-wise fashion.
- Degree of crosslinking -- a crosslinked polymer has chemical linkages between chains. In the presence of solvent, it usually swells but does not dissolve. The amount a polymer is swelled by a solvent is inversely proportional to its crosslink density, i.e., the more highly a material is crosslinked, the less it can swell.
- Ionic Character -- Many water soluble polymers are polyelectrolytes. Their ionic character depends on the number of charged groups, charge type, and charge distribution.
  - Polyelectrolytes -- macromolecules bearing a net charge
    - anionic -- macromolecules bearing a negative charge.
    - cationic -- macromolecules bearing a positive charge.
  - Polyampholytes -- macromolecules containing both cationic and anionic repeat units dispersed along the same polymer chain. These polymers may be either neutral, having the same number of cationic as anionic repeat units, or have a net charge of one sign.
    - Polybetaines -- macromolecules containing both cationic and anionic charges on the same pendent group.
- Degree of Chemical Modification – Synthetic macromolecules and many natural polymers (e.g., polysaccharides) can be chemically modified to adapt their properties to the needs of a particular application. For example, the ionic character of cellulose may be chemically modified from nonionic to anionic through nitration.

Origins of Specialty Polymers

Specialty polymers are typically water-soluble polymers that come from three origins, natural, semisynthetic, and synthetic macromolecules. Natural polymers are plant or animal based materials such as cellulose and proteins. Semisynthetic polymers are modified natural polymers manufactured by chemical derivatization of natural organic macromolecules, generally based on polysaccharides. Synthetic polymers are obtained by the polymerization of monomers synthesized from petroleum or natural gas precursors.

Natural and Semisynthetic Polymers

A significant number of water-soluble polymers are derived from biological sources, biopolymers (2). Biopolymers are an abundant and diverse class of polymers that includes polysaccharides (3), polynucleotides (4), and proteins (5). Since these polymers perform unique biological functions, they have specific microstructures and are often monodisperse.

Polysaccharides are a diverse class of biological macromolecules with a wide range of structural and behavioral characteristics (3). They are biodegradable, cyclolinear, polyhydroxyl compounds that are widely used in industry. Industrial polysaccharides have traditionally been extracted from renewable resources like starch and gums from plant seeds.

Solution properties, such as solubility, viscosity, and phase behavior, are highly dependent on the macrostructure of the chain and the chemical microstructure of the repeat units. The presence of acidic or basic functionality causes pH, electrolyte, and temperature-dependent behavior. The unique behavior of polysaccharides includes the ability to form hydrogels and lyotropic liquid crystals. These properties are largely due to hydrogen bonding and the intramolecular and/or intermolecular association of hydrophobic groups. The hydrophobic association has been shown to significantly modify the rheology of a system and is widely applied to many commercial materials such as paints, inks, personal care, and pharmaceuticals.

Polynucleotides are biopolymers that carry genetic information involved in the processes of replication and protein synthesis (4). An essentially infinite number, n^{20}, of proteins can be made by assembling the 20 amino acids in various microstructural combinations and sequence lengths. The 20 amino acids contain four major types of side chains, i.e., hydrophobic, hydrophilic, basic and acidic. Most polypeptides and proteins are water-soluble or water-swellable. The solubility of proteins varies considerably based on composition and condition of ionic strength, pH, and concentration. Those with the highest density of polar groups or electrolyte character are the most soluble. Therefore, solubility in water is lowest at the isoelectric point and increases with increasing basicity or acidity.

Polynucleotides are utilized extensively in medical, industrial, and agricultural applications. The development of recombinant DNA techniques has led to the ability to clone genes and has facilitated the production of a large numbers of proteins with significant commercial potential (6). Among the first genetically engineered proteins are insulin, the pituitary growth hormone, and interferon. Other water-soluble proteins are isolated from biological sources in a more traditional manner for a number of commercial applications. Enzymes are used as detergent additives to hydrolyze polysaccharides and proteins, to isomerize various glucose and sucrose precursors, and for mineral recovery. Supported enzymes are becoming commercially significant for large-scale substrate conversion of macromolecules(7).

Commercial Advantages of Natural and Semisynthetic Polymers
Natural and semisynthetic polymers have some commercial advantages over many synthetics, particularly in food applications.

- FDA Status -- Many have been assigned "generally recognized as safe" (GRAS) status by the US Food and Drug Administration (FDA).
- Ease of Production –
  - The processes for producing natural polymers are frequently simpler, involving harvest and refinement through chemical and mechanical operations.
  - Semisynthetics involve chemically derivatizing natural macromolecules instead of complex polymerization of monomers.
  - The facilities for producing them, therefore, are less capital-intensive, and the equipment can be more flexible.

Synthetic Polymers
Synthetic specialty polymers are obtained by the polymerization of monomers synthesized from petroleum or natural gas precursors. Linear or branched macromolecules may be formed from one or many monomers. Distribution of monomers, along the backbone or side chain, can be controlled in a number of ways, including: controlling monomer reactivity, concentration, addition order, and reaction conditions.

Major commercial synthetic specialty polymers are made by chain-growth polymerization of functionalized vinyl monomers, carbonyl monomers, or strained ring compounds. Depending on monomer structure, the polymerization may be initiated free radically, anionically, or cationically. Copolymers or terpolymers with random, alternating, block, or graft sequences can be prepared under appropriate reaction conditions. There are numerous methods used to prepare specialty polymers in the research laboratory. However, only a few are of commercial interest. Of particular commercial interest is synthesis of specialty polymers in solutions, dispersions, suspensions, or emulsions.

Commercial Advantages of Synthetic Polymers
Synthetic, semisynthetic, and natural polymers frequently can perform similar functions. However, synthetic polymers have a number of inherent advantages and are preferred in many applications for a number of reasons, including:

- Greater Flexibility – Synthetic polymers can be designed at the molecular level and are frequently used as model compounds to develop structure-property relationships for the more complex natural polymers.
- Greater Versatility – Synthetic polymers can be tailored to provide a specific property or properties for a given application. The naturals and semisynthetics are limited in the types of chemical modification because of the fragile polysaccharide backbone.
- Greater Efficacy – Since synthetic polymers can be tailored for a specific property, significantly less synthetic polymers are needed to facilitate the same performance as natural polymers.

- Lower Biological Oxygen Demand (BOD) – Since synthetic polymers have lower BOD, the effluents containing synthetic polymers are easier to treat than the other types.
- Greater Product Consistency – Since the raw materials and the reaction conditions can be well defined and controlled, synthetic polymers can be manufactured with more consistent quality than natural or semisynthetic polymers.
- Greater Price Control – Synthetics are less subject to variations in price than natural polymers due to availability.

Functionality of Specialty Polymers

Specialty polymers are best known for their built in functionality and water-solubility. This functionality is present in natural and synthetic polymers and can be broken down into two broad categories nonionic and ionic polymers. These two categories will be discussed in greater detail.

Nonionic Polymers

Nonionic specialty polymers contain diverse functional groups that do not bear a charge, such as, vinyl esters, acrylates, acrylamides, imines, and ethers. Most of these commercially important polymers are soluble in water. Their water solubility is the result of a high number of polar or hydrogen-bonding functional groups per repeat unit. Nonionic polymers are typically formed from free-radical polymerization of vinyl monomers or ring opening polymerization of strained ring compounds. An example of each and their commercial applications are given below.

Polyacrylamide is formed from the free-radical polymerization of acrylamide. Acrylamide is unique among vinyl monomers because it can be polymerized to ultrahigh molecular weight (8). Although polyacrylamide dissolves slowly, it is soluble in water in all proportions. Since polyacrylamide can be polymerized to very high molecular weight, it is a highly efficient viscosifier. Applications of polyacrylamides include flocculants, rheology control agents, and adhesives. High molecular weight copolymers of acrylamide are the most widely used polymer for water treatment. Approximately several hundred million pounds of polyacrylamide is used annually in water treatment, with a market value of approximately one billion dollars (9).

Poly(ethylene oxide) is prepared by ring-opening polymerization of ethylene oxide. It is a white free-flowing powder with commercial grades ranging from 100,000 to 5,000,000 molecular weight (10). Poly(ethylene oxide)s are completely soluble in water at room temperature, but show a lower critical solution temperature (LCST) near the boiling point of water. The solution properties of poly(ethylene oxide)s have been extensively used in commercial applications for rheology control. The unique phase transition properties are being explored for control-release in thermally-responsive applications.

Ionic Polymers

Polymers possessing ionic groups pendent to the backbone are perhaps the most important class of macromolecules, ranging from biopolymers such as polynucleotides and proteins to technologically important rheology control agents and polysoaps.

These ion-containing polymers may be divided into two groups, polyelectrolytes and polyampholytes. Polyelectrolytes have either anionic or cationic groups along the chain while the polyampholytes have both anionic and cationic groups present. Both have high charge densities and typically are water-soluble.

Water-soluble ionic polymers share a number of common properties with water-soluble nonionics, e.g., they both can act as viscosifiers. However, differences arise from the presence of charge on the macromolecular backbone and from the electrostatic interactions of mobile counterions. These differences have a significant impact on the structure of ionic polymers in solution and will be discussed further.

Polyelectrolytes

Polyelectrolytes are macromolecules bearing a net charge. Interest in these unique materials spans numerous and diverse areas. They have been investigated because of their essential participation in biological systems, such as, conformation transition of DNA (11). In ion-exchange membranes, they have been used for ion selectivity, such as recovery and separation of precious and heavy metals. Also they have been explored in polyion gels for controlled release and phase transitions, such as, drug delivery (12). Their molecular structure can be tailored to allow large conformational changes with pH, temperature, or added electrolytes. Molecular parameters that significantly influence behavior include: hydrophobic/hydrophilic balance, molecular weight, number, type and distribution of charge on the macromolecular backbone, distance of charged moiety from the backbone, and counterion type (13). Solution properties including phase behavior, hydrodynamic volume, and binding can be altered, offering utilization in flocculation, adhesion, stabilization, compatabilization, viscosification, and suspension.

Chain conformation and solubility depend on the extent of ionization and interactions with water. A characteristic feature of polyelectrolytes is the ability to attain large hydrodynamic volumes in deionized water at low concentrations (14). This effect is caused by coulombic repulsion between charged groups along the polymer chain that forces the chain into a rod-like conformation. In the presence of salt, coulombic repulsions are shielded, allowing the polymer chain to assume a more random, entropically favored conformation with a subsequent decrease in hydrodynamic volume. Electrostatic repulsions not only cause an increase in hydrodynamic volume but also increase shear sensitivity or non-Newtonian behavior.

The extent of ionization of polyelectrolytes depends on the relative base or acid strength, degree of solvation, and dielectric constant of the solvent. The structure and properties of acidic (anionic) and basic (cationic) polyelectrolytes, are discussed further.

Polyelectrolytes (Anionic) are macromolecules bearing a negative charge. Depending on the strength of the acid, they can strongly interact with positive charges. Anionic polyelectrolytes that have been studied extensively have primarily carboxylic or sulfonic acid functionality, including: poly(acrylic acid), poly(vinylsulfonic acid), poly(styrenesulfonic acid), poly(2-acrylamido-2-methylpropanesulfonic acid), poly(methacrylic acid), and their salts. Applications of anionic polyelectrolytes are diverse ranging from gels for medical applications to latex stabilizers and dye receptors in synthetic fibers.

Polyelectrolytes (Cationic) are macromolecules bearing a positive charge. Cationic functional groups can strongly interact with suspended, negatively charged particles or oil droplets. Polymers containing cationic charges can be segregated into three main categories -- ammonium, sulfonium, and phosphonium quaternaries. The commercially significant ammonium polymers are quaternary polyacrylamides, polyamines, and polyimines, including polyvinylammonium polymers. Cationic polyelectrolytes are of great importance in many industrial applications. They are used as flocculants in the clarification of drinking water and in the clean up of industrial wastes, sewage, and sludges, and as retention aids in the papermaking industry.

Polyampholytes

Polyampholytes represent a special class of polyions that contain both positively and negatively charged units on the same macromolecule. They can be polymeric zwitterions with positive and negative charges off the same backbone, or they can be polybetaines having both charges on a pendent group (15). By definition, polybetaines have an equal number of anionic and cationic charges. Whereas polyampholytes can be either neutral, having the same number of negative as positive repeat units, or have a net charge of one sign. Polyampholytes are synthesized via free-radical polymerization of anionic and cationic monomers. Due to the electron donor-acceptor nature of some cationic and anionic monomer pairs, they are prone to spontaneous polymerization. Ampholytic monomers can also be copolymerized with other water-soluble monomers.

The presence of ionic groups of opposite charges in the same macromolecule leads to unusual phase-behavior and solution properties that are largely controlled by electrostatic interactions. Unlike polyelectrolytes, polybetaines and neutral polyampholytes are frequently more soluble and show higher viscosities in salt than in deionized water. Polyampholytes solution properties are governed by coulombic attractions between anionic and cationic mer units. As the molar ratio of anionic to cationic species approaches one, coulombic interactions lead to globule-like conformations and, in many cases, insolubility in deionized water. These attractive interactions may be screened by the addition of electrolytes or change in pH, which induces a transition to a random coil conformation, often facilitating solubility. This behavior, known as the antipolyelectrolyte effect, leads to an enhancement in viscosity in the presence of electrolytes (16). These interactions can be monitored by several external parameters including number and nature of the ionic sites, polymer microstructure, solvent type, pH, and ionic strength. The overall chain conformation that results from the competition between repulsion that tends to stretch the chain (polyelectrolyte effect) and the attraction that tends to collapse it (polyampholyte effect) is highly sensitive to these factors.

Although polyampholytes have not received extensive commercial use, their dual-charge nature facilitates some unique physico-chemical properties that offer diverse application opportunities. Polyampholytes antipolyelectrolyte effect favors their use in high electrolyte solutions, such as biological and seawater applications. For example polyampholytes can be used industrially as thickeners in brine solution, in flocculation and in oil recovery processes. Because of their ability to bind to low molecular weight substances, i.e., metal ions, surfactants, dyes, drugs; polyampholytes can be used as selective chelating agents, as pigment-retention aids, and in paper manufacturing. Their intramolecular associations as a function of pH may be utilized in drug delivery. As we learn more about the unique properties of polyampholytes and use them, it is clear that their applications will continue to expand.

Applications of Specialty Polymers and Areas of Further Research

In general, ionic polymers are widely used in various fields of industry, agriculture, medicine, biotechnology, and electronics, as flocculants, coagulants, structurization agents, prolongers of drugs, biocatalysts, and sensors. With the help of gels, membranes, and films of polyelectolytes, it is possible to regulate the water regime in the ground, to purify waste-water, to disinfect ground water from radionuclides, and to create artificial nutrient media and muscles, as well as chemomechanical devices.

Specialty polymers will continue to be an area of great scientific and technological interest, due to their relevance in molecular organization in biological systems and also to various practical applications (2). These functional polymers provide unique properties to meet many of today's demanding hi-tech requirements by facilitating stimuli-responsive technologies or "smart technologies." "Smart technologies" have the ability to respond to an external stimuli in a desired fashion by altering the chemical, physical, or electrical properties of the system. This response happens in real-time and is often reversible. Examples of desired stimuli-responsive phenomena that are useful for industrial and medical applications, include the ability to respond to changes in temperature, pH, salt concentration, humidity and other environmental changes.

The key to improved specialty polymers is employing a multidisciplinary approach to develop a fundamental understanding of the underlying structure-property relationships. Synthetic specialty polymers provide a unique opportunity to systematically vary the structure of functionalized macromolecules. The ability to make model compounds of complex macromolecules can facilitate improved understanding of the underlying biological mechanisms. This will enable researchers to design, synthesize, and formulate specialty polymers with desired properties for medical and industrial applications.

References

1. Will, R.; Zaunich, J.; Debov, A.; Iskikaw, Y.; Fink, U. *Water-Soluble Polymers; SCUP Report*, SRI International, December 1998.
2. McCormick, C. L.; Bock, J.: Schulz, D. N. Encyclopedia of Polymer Science and Engineering; John Wiley: New York, NY 1989; Vol. 17, pp. 730.
3. Kennedy J. F.; White, C. A.; in Haslam, E. *Comprehensive Organic Chemistry*, Vol. 5, Pergamon Press, Oxford, UK, 1979.
4. Pitha, J. *Advances in Polymer Science*, Springer-Verlag: Berlin and Heidelberg, 1983, Vol. 50, pp. 1.
5. Creighton, T. E. *Proteins: Structures and Molecular Properties*, W. H. Freeman, New York, 1983.
6. Freifelder, D. *Recombinant DNA,* W. H. Freeman and Co., San Francisco 1977.
7. Breslow, R. *Acc. Chem. Res.* **1991**, *24*, 159.
8. Kulicke, W. M., Kniewske, R.; Klein, J. *Prog. Poly. Sci.* **1982**, *8*, 873.
9. Peaff, G.: *Chem. Eng. News*, **November 14, 1994**, 15.
10. Davidson, R. L. Handbook of Water-Soluble Gums and Resins, McGraw-Hill Inc., New York, 1980.

11. Nilsson, S.; Piculell, L. *Macromolecules* **1991**, *24*, 3804.
12. Satoh, M.; Komiyama, J. Polymeric Materials Encyclopedia; CRC Press, Inc.: Boca Raton, Florida, Vol. 8, pp. 5807.
13. Yusa, S.; Kamachi, M.; Morishima, Y. *Langmuir* **1998**, *14*, 6059.
14. Morawetz, H. Macromolecules in Solution, 2nd ed., John Wiley & Sons, Inc., New York, 1975.
15. McCormick, C. L.; Kathmann, E. Polymeric Materials Encyclopedia; CRC Press, Inc.: Boca Raton, Florida, Vol. 7, pp. 5462.
16. Higgs, P. G.; Joanny, J. F. *J. Chem. Phys* **1991**, *94 (2),* 1543.

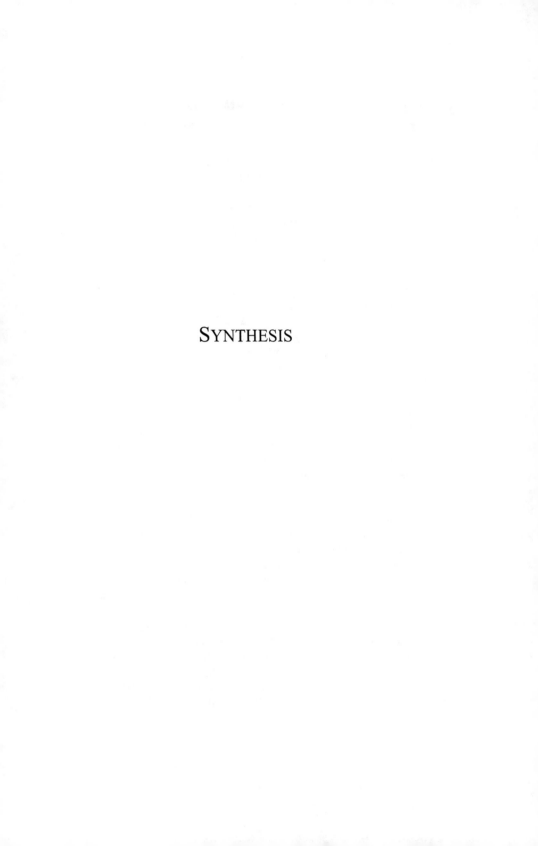

SYNTHESIS

Chapter 2

Synthesis and Aqueous Solution Behavior of Novel Polyampholytes Containing Zwitterionic Monomers

3-[(2-Acrylamido-2-methyl propyl)dimethylammonio]-1-propanesulfonate or 3-(*N*,*N*-Diallyl-*N*-methylammonio)-1-propanesulfonate

R. Scott Armentrout, Erich E. Kathmann, and Charles L. McCormick[1]

Department of Polymer Science, The University of Southern Mississippi, Hattiesburg, MS 39406

The free radical polymerizations of 3-[(2-Acrylamido-2-methyl propyl)dimethylammonio]-1-propanesulfonate with acrylamide and/or acrylic acid, and 3-(*N*,*N*-diallyl-*N*-methylammonio)-1-propanesulfonate with *N*,*N*-diallyl-*N*,*N*-dimethyl ammonium chloride or *N*,*N*-diallyl-*N*,-methyl amine have been studied. Reactivity ratios indicate random incorporation of comonomers. Molecular weights range from 3.0×10^6 to 15.1×10^6 g mol^{-1} for the acrylamido-based copolymers and from 2.2×10^4 to 6.0×10^4 g mol^{-1} for the cyclocopolymers. Second virial coefficients and viscosities decrease as sulfobetaine content increases for each of the copolymers. A transition from polyelectrolyte to polyampholyte behavior is observed with added NaCl for those copolymers with sulfobetaine monomer incorporation greater than 40 mol%

Water-soluble polymers possessing ionic groups along or pendent to the backbone are one of the most important classes of macromolecules, ranging from biopolymers such as polynucleotides to commercially important viscosifiers and polysoaps. These polymers are often classified into two groups: polyelectrolytes with *either* anionic or cationic functionality and polyampholytes that possess *both*.

A characteristic that distinguishes polyelectrolytes from polyampholytes is behavior in aqueous media. Hydrophilic polyelectrolytes exhibit extended conformations in water at low concentrations due to repulsive coulombic interactions and the associated osmotic effects. A reduction in charge by pH adjustment or addition of electrolyte allows the chain to assume a less extended, random coil conformation as evidenced by a decrease in hydrodynamic volume. In contrast, structure-behavioral relationships of hydrophilic polyampholytes are governed by coulombic attractions between anionic and cationic functional

[1]Corresponding author.

groups. As molar ratios of these groups approach one, globule-like characteristics are obtained. In cases of high charge density, phase separation from aqueous solution is typically observed. These attractive interactions can be disrupted by addition of electrolytes or pH adjustment resulting in a globule-to-coil conformational change and solubility. This behavior is referred to as the "antipolyelectrolyte effect" and is evidenced by increases in hydrodynamic volume and the second virial coefficient. Many reviews of synthetic and theoretical research on polyampholytes have appeared in the literature (*1-3*).

Our research group has been intensively studying the behavior of polyampholytes in aqueous media of moderate electrolyte content for applications in petroleum recovery, drag reduction, water remediation, and formulation of pharmaceuticals, coatings, and cosmetics. For each application, selection of appropriate comonomers and synthetic techniques can lead to desired conformational behavior under external conditions of pH, temperature, shear stress, and electrolyte concentration.

Salt-tolerant polyampholytes with potential for viscosity maintenance (or increase) at low concentration in the presence of simple electrolytes such as NaCl include polymers formed by equamolar incorporation of sulfonate and quaternary ammonium mer units (Type A) or those formed by the copolymerization of a zwitterionic sulfobetaine monomer with a water-soluble monomer (Type B) as shown in **Scheme 1**. Usually, a water-soluble mer unit, W, is included for adequate hydration. Key features of polyampholytes from sulfonated quaternary ammonium monomers (Types A and B) are discussed here, although corresponding carboxylate (*4-9*), phosphonate (*10-14*), or tertiary ammonium derivatives have been synthesized and are responsive to pH.

Type A copolymers and terpolymers have been prepared by copolymerizing vinyl pyridinium halides with alkali metal salts of sulfonate comonomers including vinylsulfonate, 2-acrylamido-2-methyl propane sulfonate, and *p*-styrene sulfonate (*15-19*). Methacrylamidopropyl-trimethyl ammonium chloride and *p*-styrene sulfonate have been terpolymerized with the hydrophilic monomer acrylamide (*20,21*). Type A copolymers and terpolymers have also been prepared from microemulsions of sodium 2-acrylamido-2-methyl-1-propanesulfonate and [2-(methacryloyloxy)ethyl]trimethylammonium chloride (*22-25*).

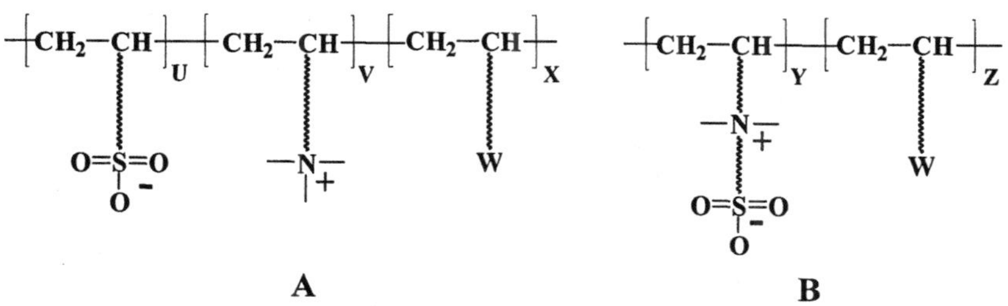

Scheme 1. Architectural types of polyampholytes.

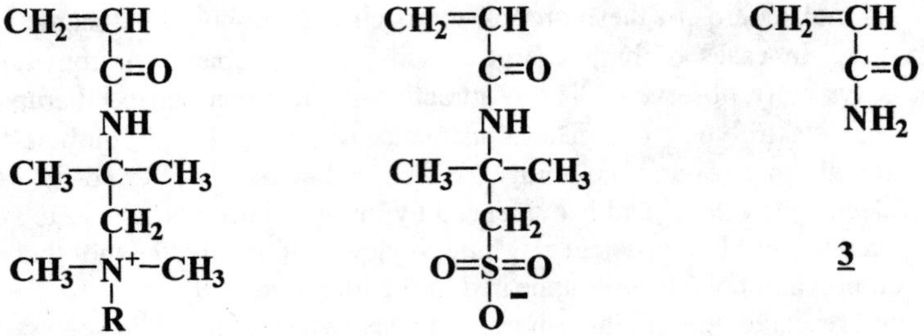

1a R=H
1b R=CH_3

2

3

Scheme 2. Example of monomers employed in the synthesis of a Type A polyampholyte.

Our group has examined copolymers and terpolymers from the hydrolytically resistant comonomers 1a, 1b, and 2 in which the cation and anion are evenly spaced from the macromolecular backbone (**Scheme 2**) (*26-30*). At equimolar incorporation of the monomers into the copolymer, a minimum in the apparent viscosity in water and a maximum in apparent viscosity in 0.514 *M* NaCl was observed – consistent with the so-called "antipolyelectrolyte effect." Similar effects are observed when acrylamide (3) is incorporated as a termonomer. A three-fold enhancement in viscosity was observed at a molar ratio of 11:13:76 for monomers (1a:2:3) in 0.5 *M* NaCl as compared to deionized water (*29*).

Polyzwitterions (Type B) containing the sulfonate functionality have been thoroughly studied beginning with the pioneering work of Hart and Timerman (*31*). In that work, zwitterionic monomers were prepared by the reaction of 2-and 4-vinylpyridine with 1,4-butanesultone. Polysulfobetaines are typically insoluble in deionized water and require a relatively high content of hydrophilic comonomer or the addition of a critical concentration of electrolyte to achieve solubility and viscosity enhancement. Polysulfobetaines have also been synthesized from acrylic (*32-36*), acrylamido (*37*), and vinyl imidazolium (*38,39*) monomers.

In this report, we describe four series of polysulfobetaines. Two systems are based upon acrylamido repeat units, while two are based upon cyclopolymers synthesized from diallyl ammonium salts. Solution behavior is discussed as functions of polymer concentration, polymer composition, concentration of added electrolytes, and pH. From these systematic investigations, it is demonstrated that the conformation of a polymer in dilute solution may be controlled by the careful design of polymer architecture, and the adjustment of environmental conditions such as NaCl concentration and pH.

Experimental

Monomer Synthesis. Monomers discussed in this work are illustrated in **Schemes 3** and **4**. Acrylamide (**3**), acrylic acid (**4**) and *N,N*-diallyl-*N,N*-dimethyl ammonium chloride (**6**) were purchased from Aldrich and purified before use. 3-[(2-Acrylamido-2-methylpropyl)dimethylammonio]-1-propanesulfonate (**5**) (*40*), *N-N*-diallyl-*N*-methyl amine (**7**) (*41*), and 3-(*N,N*-diallyl-*N*-methyl ammonio) propane sulfonate (**8**) (*42*) were prepared as previously reported.

Polymer Synthesis. *Acrylamide-based Polymers.* Acrylamide-based polymers were prepared in an 0.5 M NaCl aqueous solution at 30°C using KPS as the free radical initiator. Total monomer concentration was held constant at 0.45 M and the pH adjusted so that all monomers were in the ionized form. Reactions were terminated at <60% conversion by precipitation in acetone. The polymers were further purified by dialysis against deionized water. In the following discussion, the number appended to the polymer series is representative of the mol% of the respective sulfobetaine incorporated into the polymer (See **Tables 1, 2, 3,** and **4**).

 Cyclopolymers. Cyclopolymers were prepared in 0.5 M NaCl aqueous solution at 35°C using the photoinitiator, 2-hydroxy-1-[4-(hydroxy-ethoxy)phenyl]-2-methyl-1-propanone (Irgacure 2959) (Ciba). Total monomer concentration was held constant at 2.5 M and the pH adjusted so that all monomers were in the ionized form (pH ~ 4.0 for polymerizations of monomer **7**). Reactions were usually terminated at <50% conversion. The polymers were purified by dialysis against deionized water.

Polymer Characterization. Solution studies were performed with a Contraves LS-30 rheometer. Intrinsic viscosities were evaluated using the Huggins plot. Gated-decoupled ^{13}C NMR (Bruker AC-300) was used to determine copolymer composition. Molecular weight studies were performed in 1.0 M NaCl with a Chromatix KMX-6 low-angle light scattering spectrometer (acrylamido-based polymers) or a Brookhaven Instruments 128-channel BI-2030 AT digital correlator using a Spectra Physics He-Ne Laser operating at 632.8 nm (cyclopolymers). Refractive index measurements were carried out using a Chromatix KMX-16 differential refractometer.

Results and Discussion

Acrylamide-based polysulfobetaines. The first sulfobetaine series (**I**) consists of the copolymers of **3** and **5** (**Scheme 5**) (*40*). Compositional analysis is deteailed in **Table 1**. Reactivity ratios for comonomers **3** and **5** determined using the method of Kelen and Tüdös indicate random incorporation of the comonomers with a slight alternating tendency (r_1=0.79, r_2=0.75, r_1r_2=0.60). Weight-average molecular weights for this series range from 3.0 x 10^6 to 21.5 x 10^6 g mol^{-1}. Second virial coefficients decrease from 2.67 x 10^{-4} to 0.21 x 10^{-4} ml•mol g^{-2} as the amount of the sulfobetaine incorporated into the copolymer increases.

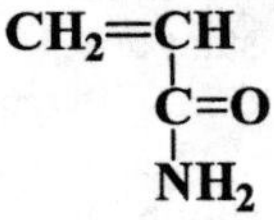

CH₂=CH
|
C=O
|
NH
|
CH₃—C—CH₃
|
CH₂
|
CH₃—N⁺—CH₃
|
CH₂
|
CH₂
|
CH₂
|
O=S=O
|
O⁻

CH₂=CH
|
C=O
|
NH₂

3

CH₂=CH
|
C=O
|
O⁻Na⁺

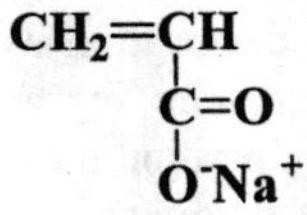

4

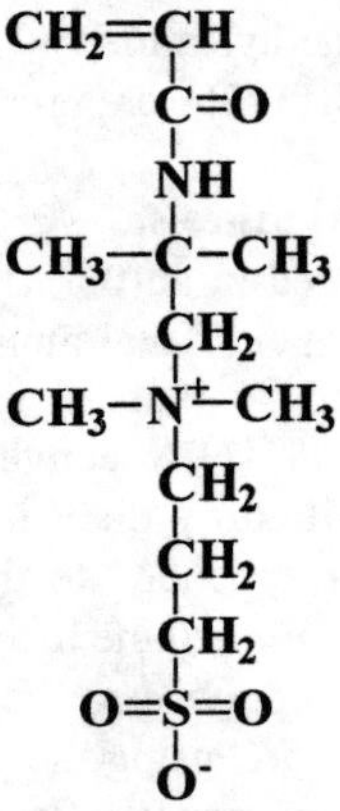

5

Scheme 3. Monomers employed in the synthesis of acrylamido-based polysulfobetaines.

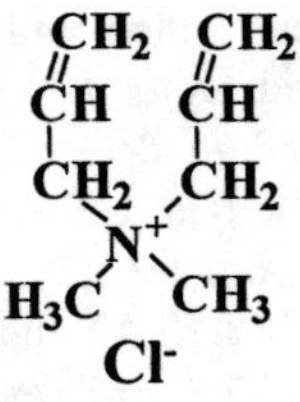

CH₂ CH₂
‖   ‖
CH   CH
| |
CH₂ CH₂
 \   /
  N⁺
 /   \
H₃C   CH₃
   Cl⁻

6

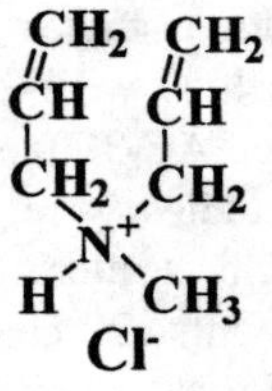

CH₂ CH₂
‖   ‖
CH   CH
| |
CH₂ CH₂
 \   /
  N⁺
 /   \
H   CH₃
   Cl⁻

7

CH₂ CH₂
‖   ‖
CH   CH
| |
CH₂ CH₂
 \   /
  N⁺
 /   \
H₃C   CH₂
       |
      CH₂
       |
      CH₂
       |
     O=S=O
       |
       O⁻

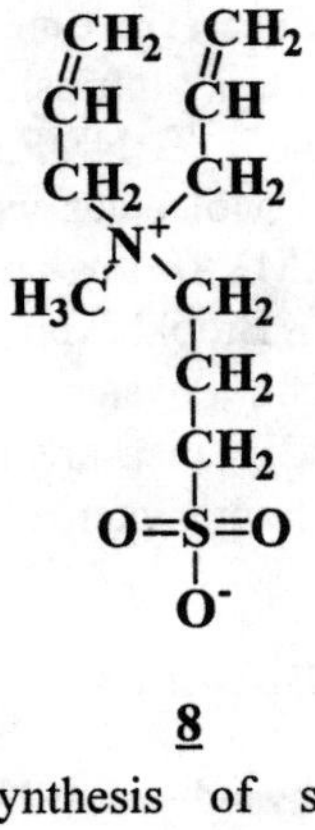

8

Scheme 4. Monomers employed in the synthesis of sulfobetaine cyclopolymers.

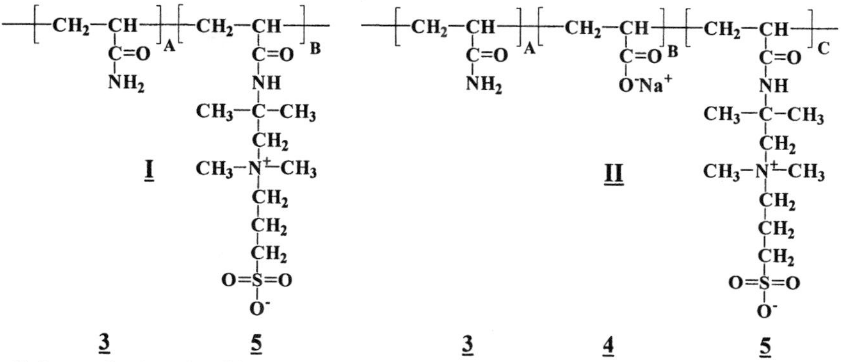

$$\left[CH_2-CH\right]_A\left[CH_2-CH\right]_B \qquad \left[CH_2-CH\right]_A\left[CH_2-CH\right]_B\left[CH_2-CH\right]_C$$

I (structure, left): mer A: C=O, NH₂ (3); mer B: C=O, NH, CH₃–C–CH₃, CH₂, CH₃–N⁺–CH₃, CH₂, CH₂, CH₂, O=S=O, O⁻ (5)

II (structure, right): mer A: C=O, NH₂ (3); mer B: C=O, O⁻Na⁺ (4); mer C: C=O, NH, CH₃–C–CH₃, CH₂, CH₃–N⁺–CH₃, CH₂, CH₂, CH₂, O=S=O, O⁻ (5)

Scheme 5. Acrylamide-based polysulfobetaines.

Table 1. Compositional analysis of sulfobetaine series I

Sample Number	Feed Ratio 3:5	5 in copolymer (mol%)[a]	M_w ($\times 10^{-6}$ g mol^{-1})[c]	A_2 ($\times 10^4$ ml mol g^{-2})[c]
I- 10	90:10	10.0	7.0	1.49
I- 26	75:25	26.0	8.2	1.33
I- 38	60:40	38.3	15.1	0.75
I- 57	40:60	57.3	5.4	0.49
I- 69	25:75	69.2	6.2	0.48
I-100	0:100	100[b]	3.0	0.25

[a] Determined from ^{13}C NMR [b] Theoretical value [c] Determined in 1 M NaCl

Figure 1 illustrates the effect of NaCl concentration on intrinsic viscosities for copolymers of series I with incrementally increasing composition of sulfobetaine mer units. Copolymers with 10, 26, and 38 mole percent of 5 (I-10, I-26, and I-38) are soluble in deionized water and show increases in viscosity across the NaCl concentration range. I-10 and I-28 may be aggregated (multimers) in deionized water, explaining the initial decrease upon addition of salt. The homopolymer of 5 (I-100) and copolymers I-57 and I-69 are insoluble in deionized water and require a critical concentration of NaCl for dissolution. In deionized water, the intramolecular dipole-dipole interactions in these copolymers dominate, allowing little hydration. Addition of electrolyte induces the globule-to-random coil transition typical of polyampholyte hydration.

In an extension of the above studies, terpolymers of 3, 4, and 5 (series II) (**Scheme 5**) were synthesized (37). Compositional analysis for this series of terpolymers is listed in **Table 2**. Weight-average molecular weights for this series range from 3.0 x 10⁶ to 7.9 x 10⁶ g mol⁻¹. It was initially postulated that these polymers would exhibit polyzwitterionic or polyelectrolyte behavior depending upon the pH. However, at low pH values, hydrogen bonding between the carboxylic acid and amide moieties limits solubility even in the presence of salts. Therefore, all solution studies were conducted at pH = 8. Under these conditions,

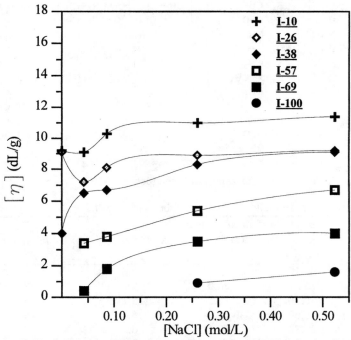

Figure 1. Intrinsic viscosities of the copolymers of <u>3</u> and <u>5</u> (series <u>I</u>) at 25°C as a function of NaCl concentration (γ=5.96 s⁻¹).

Table 2. Compositional analysis of sulfobetaine series <u>II</u>

Sample Number	Feed Ratio 3: 4: 5	4:5 in copolymer (mol%)ᵃ	M_w (x10⁻⁶ g mol⁻¹)ᶜ	A_2 (x10⁴ ml mol g⁻²)ᶜ
<u>II- 3: 5</u>	90: 5: 5	3.2: 5.0	3.0	2.70
<u>II- 5:10</u>	80:10:10	4.6:10.2	5.4	2.95
<u>II-13:31</u>	50:25:25	13 :31	7.9	2.23
<u>II-25:50</u>	20:40:40	25 :50	4.7	2.23
<u>II- 6: 0</u>	90:10: 0	5.8: 0	4.0	5.76

ᵃ Determined from ¹³C NMR ᵇ Theoretical value ᶜ Determined in 1 *M* NaCl

polyelectrolyte behavior is observed for all of the systems. This is illustrated in **Figure 2**, which shows the intrinsic viscosities as a function of salt concentration for the terpolymer <u>II-5:10</u> and the copolymer <u>II-6:0</u>. At low ionic strengths, the high intrinsic viscosity of this system is due to the polyelectrolyte character of the polymer backbone. As the ionic strength is increased, the intrinsic viscosities decreased as expected. However, the zwitterionic nature of the sulfobetaine units (<u>II-5:10</u>) retards the total loss of viscosity exhibited by <u>II-6:0</u>. Clearly, a balance of the polyelectrolyte and antipolyelectrolyte effects can be attained by proper selection of constitutive monomers.

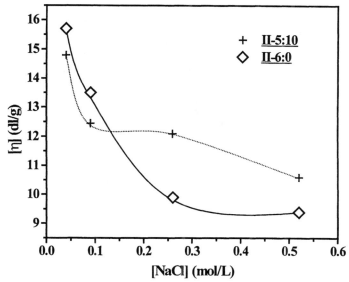

Figure 2. Intrinsic viscosities of <u>II-5:10</u> and <u>II-6:0</u> at 25°C as a function of NaCl concentration (γ=5.96 s^{-1}).

Sulfobetaine cyclopolymers. The first sulfobetaine cyclopolymer series (<u>III</u>) consists of the copolymers of <u>6</u> with <u>8</u> (**Scheme 6**) (*42,43*). Compositional analysis of this series of cyclopolymers is listed in **Table 3**. Reactivity ratios, determined via nonlinear least squares analysis[44] of chemical compositions determined from gated-decoupled ^{13}C NMR, indicate random incorporation of the comonomers (r_1=1.14 r_2=0.97). The five-membered ring structure common to polymerized diallyl ammonium salts was retained in the sulfobetaine mer unit. Weight average molecular weights for this series range from 3.04 x 10^4 to 6.03 x 10^4 g mol^{-1}. Second virial coefficients decrease from 8.78 x 10^{-4} to 2.50 x 10^{-4} ml•mol g^{-2} as the amount of <u>8</u> incorporated into the copolymer increases.

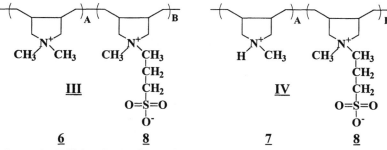

Scheme 6. Sulfobetaine cyclocopolymers.

Table 3. Compositional analysis of sulfobetaine series III.

Sample Number	Feed Ratio 6:8	8 in copolymer (mol%)[a]	M_w ($\times 10^{-4}$ g mol^{-1})[c]	A_2 ($\times 10^4$ ml mol g^{-2})[c]
III- 0	100:0	0[b]	3.04	8.78
III- 11	90:10	11	3.38	7.24
III- 23	75:25	23	4.28	6.04
III- 40	60:40	40	5.44	3.44
III- 55	40:60	55	4.97	2.50
III- 82	20:80	82	5.52	2.94
III-100	0:100	100[b]	6.03	3.18

[a] Determined from ^{13}C NMR [b] Theoretical value [c] Determined in 1 M NaCl

Figure 3 illustrates the reduced viscosities for copolymer series III as a function of added electrolyte. It was discovered that copolymers containing ≤55 mol% of 8 are soluble in deionized water due to the excess positive charge present along the polymer. As expected, classical polyelectrolyte behavior is observed upon the addition of electrolyte (i.e. a decrease in reduced viscosity). However, in the polymers with incorporations of 8 ≥40 mol% both polyelectrolyte and polyampholyte characteristics contribute to the observed behavior. The initial decrease in reduced viscosity is followed by an increase as the electrolyte concentration increases above a critical concentration. The resulting minimum in reduced viscosity corresponds to an environment in which the charge-charge repulsions along the cyclopolymer backbone are effectively screened such that the dipole-dipole intramolecular interactions dominate, leading to a collapse of the polymer coil. As the electrolyte concentration is increased above this critical concentration, dipole-dipole interactions are effectively screened and an increase in the viscosity is realized. It must be pointed out that for sample III-55 at 0.075 $M <$ [NaCl] < 0.5 M, the dipole-dipole associations are sufficiently prevalent to bring about macroscopic phase separation.

In an extension to the above work, the pH-responsive monomer, 7, has been copolymerized with 8 (series IV) (**Scheme 6**) (*45*). Compositional analysis of this series of cyclopolymers is listed in **Table 4**. Reactivity ratios indicate random incorporation of the comonomers (r_1=0.67, r_2=1.13) slightly favoring incorporation of the sulfobetaine monomer, 8. The five-membered ring structure common to polymerized diallyl ammonium salts was retained in the polymerized mer units as determined by ^{13}C NMR analysis. Weight average molecular weights for this series range from 2.27 x 10^4 to 4.01 x 10^4 g mol^{-1}.

Figure 4 illustrates the apparent viscosity for copolymer series IV as a function of polymer concentration and degree of ionization (α) of the 7 mer unit as determined by potentiometric titration with NaOH (data not shown). In the protonated form (α=1), the polymer should assume a conformation resembling that of series III. Analysis of the panels in **Figure 4**, reveals an expanded conformation at high degrees of ionization. Similarly, as the incorporation of the sulfobetaine monomer increases, the polymer chain assumes a more collapsed conformation as

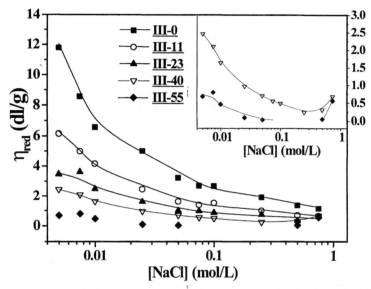

Figure 3. Reduced viscosities of the copolymers of $\underline{6}$ and $\underline{8}$ (series $\underline{III}$) at 25°C as a function of NaCl concentration ([Polymer]=0.15 g/dl, γ=5.96 s^{-1}).

Table 4. Compositional analysis of sulfobetaine series $\underline{IV}$.

Sample Number	Feed Ratio $\underline{6}$:$\underline{8}$	$\underline{8}$ in copolymer (mol%)[a]	M_w ($\times 10^{-4}$ g mol^{-1})[c]	A_2 ($\times 10^4$ ml mol g^{-2})[c]
IV- 0	100:0	0[b]	3.05	69.4
IV- 12	90:10	12	2.27	8.53
IV- 34	75:25	34	2.67	5.82
IV- 44	60:40	44	3.36	3.47
IV- 66	40:60	66	3.84	2.79
IV- 82	20:80	82	4.01	3.28
IV-100	0:100	100[b]	6.03	3.18

[a]Determined from ^{13}C NMR [b]Theoretical value [c]Determined in $1M$ NaCl, pH=6.0

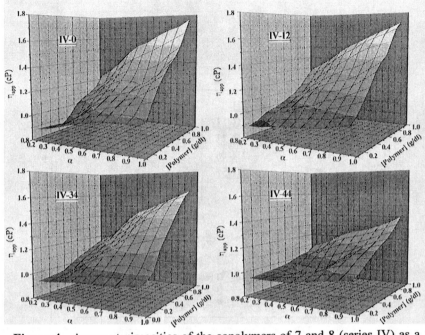

Figure 4. Apparent viscosities of the copolymers of <u>7</u> and <u>8</u> (series <u>IV</u>) as a function of polymer concentration and degree of ionization of the *N,N*-diallyl-*N*-methyl amine (<u>7</u>) mer unit.

indicated by a lower apparent viscosity. As the degree of ionization of the <u>7</u> mer unit decreases, a collapse of the polymer occurs. This collapse is governed by the transition of the <u>7</u> mer units from an ionic to a hydrophobic mer unit. (The *N-N*-diallyl-*N*-methyl amine monomer in the unprotonated form is insoluble in water.) Therefore, the collapse is governed by the formation of intramolecular hydrophobic microdomains formed by the unprotonated mer unit. At $\alpha \approx 0$ the homopolymer of <u>7</u> precipitates from solution. However, as the incorporation of the sulfobetaine increases to 12 mol% (<u>IV-12</u>) precipitation does not occur, but rather macroscopic phase separation occurs. As the sulfobetaine incorporation increases to 34 and 44 mol% (<u>IV-34</u> and <u>IV-44</u>) no phase separation occurs at low α. In these cases, the hydrophobic interactions are diminished due to the presence of the sulfobetaine moieties. Therefore, the polymer assumes a more relaxed conformation as evidenced by higher apparent viscosities at low α for <u>IV-34</u> and <u>IV-44</u> than <u>IV-0</u> and <u>IV-12</u>. It is also interesting to note that at low α, <u>IV-66</u> also becomes macroscopically phase separated. However, instead of being driven by hydrophobic interactions, this phase separation is driven by dipole-dipole interactions. Therefore, by adjusting the composition of copolymer series <u>IV</u> and the degree of ionization of the <u>7</u> mer unit, the conformation that the copolymer chain assumes in dilute solution may be controlled.

Conclusions

In this chapter, we have illustrated the effects of molecular architecture of polysulfobetaines on solution behavior under specific environmental conditions of pH, added electrolytes, and polymer concentration. The nature of the comonomer and amount of incorporation of the sulfobetaine within the polymer chain dictate the polymer solubility and solution behavior. Polyampholyte behavior is realized for acrylamide-based systems containing the sulfobetaine moiety. Polyelectrolyte behavior is coupled with polyampholyte behavior for cyclopolymers containing >40mol% sulfobetaine. Incorporation of the sulfobetaine monomer hinders hydrophobic association for the pH responsive copolymers of series IV at low degrees of ionization.

Acknowledgements

Support for this research by the Office of Naval Research, and the Department of Defense is gratefully acknowledged.

References

1. McCormick, C.L.; Bock, J.; and Schulz, D.N. Encyclopedia of Polymer Science and Enginering 2nd Edition; John Wiley and Sons, Inc.: New York, **1989** *17*, 730.
2. Bekturov, E.A.; Kudaibergenov, Rafikov, S.R. *J. Macromol. Sci.-Rev Macromol. Chem. Phys.* **1990**, *C30(2)* 233.
3. Salamone, J.C.; Rice, W.C. Encyclopedia of Polymjer Science and Engineering, 2nd Edition.; John Wiley and Sons, Inc.: New York **1988** *11*, 514.
4. Ladenheim, H.; Morawetz, H. *J. Polym. Sci.* **1957**, *26*, 251.
5. Rosenheck, K.; Katchalsky, A. *J. Polym. Sci.* **1958**, *32*, 511.
6. Wielema, T. A.; Engberts, J. B. F. N. *Eur. Polymer J.* **1988**, *24*, 647.
7. Hsu, Y. G.; Hsu, M. J.; Chen, K. M. *Makromol. Chem.* **1991**, *192*, 999.
8. Kathmann, E. E.; White, L. A.; McCormick, C. L. *Polymer* **1997**, 871.
9. Kathmann, E. E.; McCormick, C. L. *J. Poly. Sci. Part A. Poly. Chem.* **1997**, 231.
10. Yasuzawa, M.; Nakaya, T.; Imoto, M. *J. Macromol. Sci Chem. Part A* **1986**, *23(8)*, 963.
11. Nakaya, T.; Yasuzawa, M.; Imoto, M. *Macromolecules* **1989**, *22*, 3180.
12. Nakaya, T.; Yasuzawa, M.; Imoto, M. *Macromolecules* **1989**, *22*, 3180.
13. Hamaide, T.; Germanaud, L.; Le Perchec, P. *Makromol. Chem.* **1986**, *187*, 1097.
14. Pujol-Fortin, M. L.; Galin, J. C.; Morawetz, H. *Polymer* **1994**, *35*, 1462.
15. Salamone, J. C.; Watterson, A. C.; Hsu, T. D.; Tsai, C. C.; Mahmud, M. U. *J. Polym. Sci. Polym. Lett. Ed.*, **1977**, *15*, 487.
16. Salamone, J. C.; Watterson, A. C.; Hsu, T. D.; Tsai, C. C.; Mahmud, M. U; Wisniewski, A. W.; Israel, S. C. *J. Polym. Sci. Polym. Symp.*, **1978**, *64*, 229.
17. Salamone, J. C.; Quach, L.; Watterson, A. C.; Krauser, S.; Mahmud, M. U; *J. Macromol. Sci. Part A*, **1985**, *22*, 653.

18. Salamone, J. C.; Ahmed, I.; Rodriguez, E. L.; Quach, L.; Watterson, A. C.; *J. Macromol. Sci. Part A*, **1988**, *26*, 1234.
19. Salamone, J. C.; Mahmud, N. A.; Mahmud, M. U.; Watterson, A. C. *Polymer*, **1982**, *23*, 843.
20. Peiffer, D.G.; Lundberg, R.D. *Polymer*, **1985**, *26*, 1058.
21. Peiffer, D. G.; Lundberg, R. D.; Dvdevani, I. *Polymer*, **1986**, *27*, 1453.
22. Corpart, J.; Selb, J.; Candau, F. *Makromol. Chem., Makromol.Symp.,* **1992**, *53*, 253.
23. Corpart, J.; Candau, F. *Macromolecules*, **1993**, *26*, 1333.
24. Corpart, J.; Selb, J.; Candau, F. *Polymer*, **1993**, *34*, 3873.
25. Skouri, M.; Munch, J.P.; Candau, F.S.; Neyret, S.; Candau, F. *Macromolecules*, **1994**, *27*, 69.
26. McCormick, C. L.; Johnson, C. B. *Macromolecules* **1988**, *21*, 687.
27. McCormick, C. L.; Johnson, C. B. *Macromolecules* **1988**, *21*, 694.
28. McCormick, C. L.; Salazar, L. C. *Macromolecules* **1992**, *25*, 1896.
29. McCormick, C. L.; Johnson, C. B. *Polymer* **1990**, *31*, 1100.
30. McCormick, C.L.; Salazar, L.C. *Polymer* **1992**, *33*,4384.
31. Hart, R.; Timmerman, D. J. *Polymer Sci.* **1958**, *28*, 638.
32. Schulz, D. N.; Peiffer, D. G.; Agarwal, P.K.; Larabee, J.; Kaladas, J. J.; Soni, L.; Handwerker, B.; Garner, R. T.; *Polymer* **1986**, *27*, 1734.
33. Schulz, D. N.; Kitano, K.; Danik, J.A.; Kaladas, J. J. *Polym. Mat. Sci. Eng.* **1987**, *147*, 149.
34. Huglin, M.B.; Radwan, M.A. *Polymer International* **1991**, *26*, 97.
35. Konack, C.; Rathi, R. C.; Kopeckova, P.; Kopecek, J. *Polymer* **1993**, *34*, 4767.
36. Konack, C.; Rathi, R. C.; Kopeckova, P.; Kopecek, J. *Macromolecules* **1994**, *27*, 1992.
37. Kathmann, E. E.; Davis, D. D.; McCormick, C. L. *Macromolecules* **1994**, *27*, 3156.
38. Salamone, J. C.; Volksen, W.; Israel, S.C.; Olson, A. P.; Raia, D. C. *Polymer* **1977**, *18*, 1058.
39. Salamone, J. C.; Volksen, W.; Olson, A. P.; Israel, S.C. *Polymer* **1978**, *19*, 1157.
40. McCormick, C. L.; Salazar, L.C. *Polymer* **1992**, *33*, 4617.
41. Chang, Y.; McCormick, C. L. *Polymer* **1994**, *35*, 3603.
42. Armentrout, R. S.; McCormick, C. L. *Polymer Preprints;* The American Chemical Society: Washington, D.C. **1998**, *39(1)*, 617.
43. Armentrout, R. S.; McCormick, C. L. *Manuscript in preparation.*
44. Tidwell, P.W., and Mortimer, G.A. *J. Polym. Sci.: Part A* **1965**, *2*, 369.
45. Armentrout, R. S.; McCormick, C. L. *Manuscript in preparation.*

Chapter 3

Synthesis and Emulsion Copolymerizations of Styrene Sulfonate Monomers for Cross-Linked Polyampholyte Latexes

Kenneth W. Hampton and Warren T. Ford

Department of Chemistry, Oklahoma State University, Stillwater, OK 74078

p-Styrenesulfonyl chloride (SSC, **2**) was prepared with improved purity. Methyl *p*-styrenesulfonate (MSS, **3**), and methyl *m,p*-vinylbenzylsulfonate (MVBS, **9**) were prepared for the first time. To form polyampholyte latexes, the styrene sulfonate monomers (SSC, MSS, MVBS) were copolymerized in surfactant-free emulsions with styrene, vinylbenzyl chloride (VBC, **6**), divinylbenzene, and either vinylbenzyl (trimethyl)ammonium chloride (VBTMAC) or sodium *p*-styrenesulfonate (NaSS, **1**) for charge stabilization. Methyl *m,p*-vinylbenzyl poly(ethylene oxide) (MVBPEO, **5**) improved the colloidal stabilities of the latexes. All latexes containing a styrene sulfonate monomer were polydisperse.

A polyampholyte contains positively and negatively charged repeat units randomly dispersed along the same linear chain. This polymer may be either neutral, having the same number of negative as positive repeat units, or have a net charge of one sign. If the net charge is large, the polymer behaves as a polyelectrolyte: Its random coil size contracts with increasing concentrations of added salts. However, a polyampholyte with close to neutral charge can exhibit an anti-polyelectrolyte behavior: Its solubility increases and its random coil size expands with increasing concentrations of added salts (*1-10*). Thus polyampholytes could find many useful applications in high electrolyte solutions, such as seawater, which covers the majority of the surface of the earth.

Water-soluble polyampholytes have been synthesized by copolymerization of both anionic and cationic monomers into the polymer backbone (*2-8*) or by incorporating zwitterionic monomer units into the polymer (*9-10*). In pure water most polyampholytes having randomly distributed (+) and (-) functional groups, in the 40/60 to 60/40 range of mol ratios are insoluble due to intrapolymer electrostatic attractions (*6-8*). The solubility and the solution viscosity increase upon adding salt due to breakup of the intrapolymer ion aggregates and expansion of the polymer coil.

Previously in this lab, monodisperse quaternary ammonium ion (N+) latexes were synthesized (*11*) and used to catalyze reactions in aqueous dispersions (*12*). However, most conventional charged latexes (including the N+ latexes) precipitate in dispersions containing high concentrations of salts. The goal of the work described here is to synthesize cross-linked polyampholyte latexes that are stable in high concentrations of salt. The synthesis proceeds via emulsion copolymerization of vinylbenzyl chloride (VBC), divinylbenzene, styrene, and a sulfonate monomer (SSC, MSS or MVBS) to produce a precursor latex. Reaction of the precursor latex with trimethylamine (TMA) or other tertiary amines in one step produces a polyampholyte latex containing quaternary ammonium ions (N^+) and sulfonate ions (SO_3^-) as outlined in Figure 1. The compositions of the polyampholyte latexes are controlled by the amounts of functional monomers used in the copolymerization.

Experimental

General. Irganox 1010 (CIBA GEIGY, a tetra-2,6-di-t-butylphenolic inhibitor), sodium p-styrenesulfonate (NaSS, Polysciences), thionyl chloride (Aldrich), sodium iodide (J.T. Baker), sodium sulfite (Fisher) and 25 wt % trimethylamine in water (Eastman) were used as received. The monomers vinylbenzyl chloride, styrene and divinylbenzene (Aldrich) were distilled under vacuum, and filtered through alumina prior to use. (m, p-Vinylbenzyl)trimethylammonium chloride (VBTMAC) was prepared previously (*11*), and used as a 0.0302 M aqueous solution (as measured with a chloride selective electrode). The initiator, VA-044 (2,2'-azobis(N,N'-dimethyleneisobutyramidine) dihydrochloride, Wako) was used as received. N,N-dimethylformamide (DMF, Aldrich) was dried with CaH_2, and distilled onto 5Å molecular sieves. THF (Aldrich) was distilled over sodium. Poly(ethylene glycol) monomethyl ether (MPEG-OH, Aldrich, MW 2000) was dried by azeotropic distillation with toluene and under vacuum at 50 °C overnight, and stored under nitrogen. Water was deionized by an E-pure (Barnstead) 3-module system to a conductivity of 0.65 μmhos. Transmission electron micrographs were obtained and particle sizes were calculated as described elsewhere (*11*).

p-Styrenesulfonyl Chloride (2). A 1000 mL 4-neck flask equipped with a mechanical stirrer, thermometer, and nitrogen inlet was placed into a 40 °C oil bath. Irganox 1010 (1.0 g) was added with 210 mL of anhydrous DMF. NaSS (70 g, dried in vacuum at 40 °C) was added in small amounts. The NaSS was only slightly soluble. The contents of the flask were chilled in a ice bath. Chilled thionyl chloride (175 mL) was added dropwise over 1 h through an ice-jacketed addition funnel while the stirred mixture was held at <5 °C and the mixture was stirred for 1 h. The mixture was kept under nitrogen for 12 h at 5 °C, poured into 600 mL of ice water and extracted with 2x 400 mL of toluene. The extract was washed with 2x 400 mL of water, dried over anhydrous sodium sulfate, and concentrated on a rotary evaporator at <40 °C to leave 43.9 g (79.5%) of a straw-colored oil. ^{1}H NMR (400 MHz, CDCl$_3$) δ 5.55 [d, 1H, J = 10.3 Hz], 5.95 [d, 1H, J = 17.5 Hz], 6.8 [dd, 1H], 7.6-8.0 [dd, 4H, C_6H_4].

Methyl p-Styrenesulfonate (3). In 185 mL of dichloromethane and 75 mL of methanol was dissolved 37.3 g of SSC (**2**). Irganox 1010 (3 mg) was added and the mixture was stirred in an ice bath for 30 min. A solution of 27.5 g of potassium hydroxide in 31 mL of water was cooled in an ice-jacketed addition funnel and added dropwise over 45 min. The temperature was raised to 25 °C, and the mixture was stirred for 1 h. Ice water (300 g) was added, and the mixture was neutralized with dilute sulfuric acid and extracted with 3x 100 mL of dichloromethane. The extracts were combined, dried over anhydrous sodium sulfate, filtered, and rotary evaporated at <40 °C. The residue was dissolved in 5 mL of dichloromethane and filtered through 10 g of silica gel. Rotary evaporation of the solvent at <40 °C, gave

24.1 g (67%) of a yellow oil. ^{1}H NMR (400 MHz, CDCl$_3$) δ 3.8 [s, 3H, CH$_3$], 5.45 [d, 1H, J = 10.9 Hz], 5.95 [d, 1H, J = 17.7 Hz], 6.9 [dd, 1H], 7.6-7.9 [dd, 4H, C$_6$H$_4$].

Methyl *m,p*-Vinylbenzyl Poly(ethylene oxide) (5). Into a 100 mL three-necked flask were added 7.0 g (3.5 mmol) of dry MPEG-OH (4) and 35 mL of THF. The mixture was heated to reflux under nitrogen and 0.420 g (17.5 mmol) of sodium hydride (60% dispersion in mineral oil) was added. The mixture was refluxed for 20 h, and cooled to 25 °C under nitrogen. VBC (6, 2.67 g, 17.5 mmol) was added, and the mixture stirred under nitrogen for 48 h. The mixture was poured into 250 mL of ice-cold acetone under nitrogen and centrifuged. The supernatant fluid was decanted, the acetone was removed by rotary evaporation, and the residue was extracted with 100 mL of water and 3x 50 mL of dichloromethane. The organic extracts were combined, dried over anhydrous MgSO$_4$, filtered, concentrated by rotary evaporation to 30 mL, and poured into ice-cold diethyl ether under nitrogen. The product was collected by vacuum filtration and dried in vacuum to give 5.56 g (81%) of white solid. ^{1}H NMR (400 MHz, DMSO-d$_6$) δ 3.25 [s, 3H, CH$_3$O-], 3.51 [bs, -CH$_2$CH$_2$O-], 4.49 [d, 2H, ArCH$_2$O-], 5.25 [dd, 1H, J = 11.8 Hz], 5.82 [dd, 1H, J = 17.2 Hz], 6.55 [m, 1H], 7.2-7.5 [m, 4H, C$_6$H$_4$].

Sodium *m,p*-Vinylbenzylsulfonate (7). A solution of VBC (6, 30.0 g, 0.20 mol) and sodium sulfite (Na$_2$SO$_3$•7H$_2$O, 31.5 g, 0.22 mol) in acetone/water (2:1, 600 mL) was stirred magnetically for several minutes. The reaction was started by the addition of 1.525 g of sodium iodide. After stirring in a 45 °C bath for 24 h the mixture was filtered, and the retained sodium chloride was washed with acetone. The acetone/water was removed under reduced pressure to yield white crystals, which were rinsed with 100 mL dichloromethane and dried to give 35.7 g (82.6%) of a white solid. ^{1}H NMR (400 MHz, DMSO-d$_6$) δ 3.7 [d, 2H, CH$_2$], 5.2 [t, 1H, J = 10.0 Hz], 5.75 [d, 1H, J = 17.7 Hz], 6.7 [dd, 1H], 7.2-7.4 [m, 4H, C$_6$H$_4$]. The mixture was 42% *meta* and 58% *para* by ^{1}H NMR integration of the CH$_2$ region at 3.7 ppm and analysis of the multiplets at 7.2-7.4 ppm.

Methyl *m,p*-Vinylbenzylsulfonate (9). NaVBS (7) (38 g, dried in the vacuum at 40 °C) was treated with thionyl chloride in DMF as in the synthesis of SSC (2). The resulting VBSC (8) was not isolated but, after removal of the toluene by rotary evaporation was converted directly into MVBS (9) by potassium hydroxide in dichloromethane/methanol as in the synthesis of MSS (3). A yellow oil was recovered in 44.1% yield (16.2 g). ^{1}H NMR (300 MHz, CDCl$_3$) δ 3.76 [d, 3H, CH$_3$], 4.35 [d, 2H, CH$_2$], 5.3 [d, 1H, J = 10.9 Hz], 5.76 [d, 1H, J = 17.6 Hz], 6.73 [dd, 1H], 7.3-7.5 [m, 4H, C$_6$H$_4$].

General One Shot Latex Synthesis. A 100-mL three-neck round-bottom flask equipped with an overhead stirrer with a Teflon blade, a condenser, and a nitrogen inlet was charged with 70 mL of water and either 0.060 g VBTMAC in water or solid NaSS (1). The mixture was stirred and purged with nitrogen for 20 min. A mixture of styrene, VBC, SSC and DVB was added. The mixture was stirred and heated for 20 min in a 60 °C oil bath, 1 wt % of VA-044 or the amounts of potassium persulfate and sodium bisulfite required to make the aqueous phase 4.0 mM in each was added, and the mixture was stirred at 60 °C for 10 h. The latex was filtered through a cotton plug to remove a small amount of coagulum.

Semibatch Emulsion Polymerization. A 100 mL 3-neck round bottom flask was equipped with a nitrogen inlet, Teflon blade stirrer, and condenser with syringe pump inlet. A typical recipe is shown in Table III. All water and initial monomers were charged to the flask at room temperature and purged with nitrogen for 10 min while heating in a 60 °C oil bath and stirring at 500 rpm. The reaction was initiated by adding VA-044. After 20 min the second mixture of monomers was fed to the flask over 70 min by a syringe pump. Heating and stirring were continued for 2 h. The latex was filtered through a cotton plug to remove coagulum.

Quaternization and Cleaning of the Latexes. A mixture of 50 mL of latex (about 5 g solid), 25 mL of H_2O and 17.0 g of 25 wt % trimethylamine in water was added to a 150 mL beaker and sealed in a stainless steel reactor. The mixture was held under pressure with magnetic stirring for 48 h with the reactor half immersed in a 60 °C oil bath. The excess trimethylamine was evaporated by bubbling nitrogen through the mixture. A Spectra/Por® dialysis membrane with molecular weight cutoff of 50,000 was washed with deionized water several times to remove sodium azide. The latex was dialyzed against deionized water for one week with frequent changing of the water, and was ultrafiltered in a stirred filtration cell through a 0.1 μm cellulose acetate/nitrate membrane (Millipore) under 20 psi of nitrogen. Deionized water was added repeatedly for several days until approximately 800 mL of filtrate was collected, and the filtrate had a constant conductivity of 1.5 μmhos. The percent solids (w/v) of each latex was measured by evaporating 3x1 mL of the dispersion at 100 °C to constant weight.

Results and Discussion

The strategy for synthesis of cross-linked polyampholytes is in Figure 1. The monomers designed to produce negative charge in the polyampholytes and to sterically stabilize the latexes were prepared as shown in Figures 2-4 by modified literature procedures (*13,14*). The initial emulsion copolymerization is carried out with a small amount of either (vinylbenzyl) trimethylammonium chloride or NaSS to stabilize the latex. To avoid the presence of low molar mass amphiphiles in the polyampholytes, no surfactant is used. Reaction of the precursor latex with trimethylamine converts the styrenesulfonic ester or acid chloride units to the negative sulfonate ions and the vinylbenzyl chloride units to the positive quaternary ammonium ions of the polyampholyte.

Polymerizations with SSC (2). Previously NaSS-styrene copolymer latexes have been made, but <3% of NaSS could be incorporated without simultaneous formation of a soluble polyelectrolyte or a bimodal distribution of particle sizes (*15-17*). Hence, a lipophilic monomer that is freely miscible with styrene and VBC must be employed in order to achieve a high charge density styrenesulfonate latex.

Several groups have used SSC (**2**) for the preparation of ion exchange membranes (*18*) and microspheres (*19*). In both cases, hydrolysis of the sulfonyl chloride to the sulfonic acid during polymerization was not an issue. Hydrolysis of SSC during emulsion polymerization was a concern, but we proceeded because SSC is hydrolytically stable enough to be purified during preparation by washing with water.

Freshly prepared SSC was used in a series of one shot polymerizations using both cationic and anionic initiators as shown in Table I. The first entry C50N is a control experiment that demonstrates formation of a stable latex form the cationic monomer, styrene, and VBC with no sulfonate monomer. The next two entries were designed to give approximately equal amounts of (+) and (-) charges after quaternization, but both latexes were unstable at the precursor stage, perhaps because of partial hydrolysis of the sulfonyl chloride. When potassium persulfate was

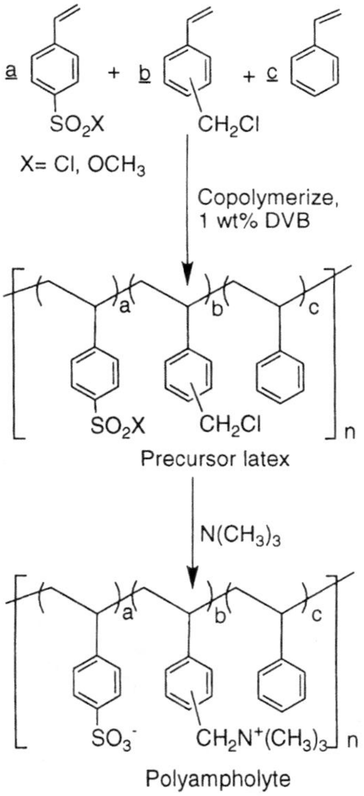

Figure 1. Synthesis of Cross-linked Polyampholyte Latexes.

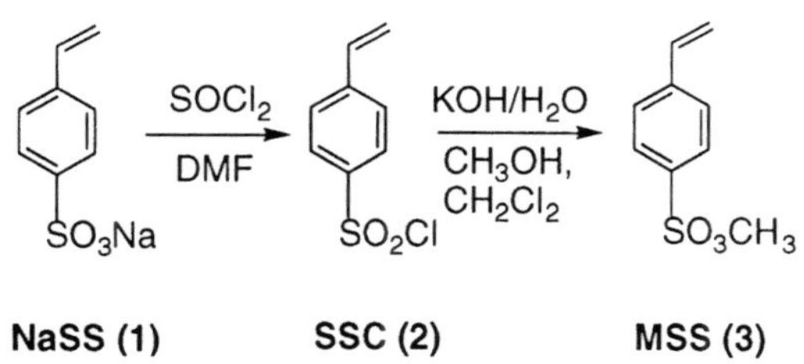

NaSS (1)   **SSC (2)**   **MSS (3)**

Figure 2. Synthesis of MSS.

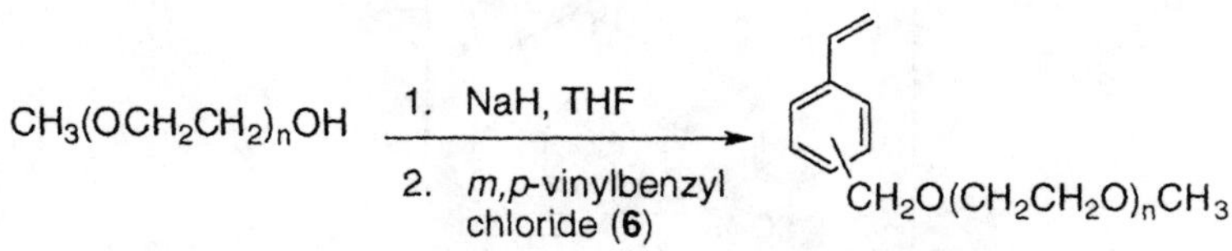
CH3(OCH2CH2)nOH
1. NaH, THF
2. m,p-vinylbenzyl chloride (6)
CH2O(CH2CH2O)nCH3

MPEG-OH (4) MVBPEO (5)

Figure 3. Synthesis of MVBPEO.

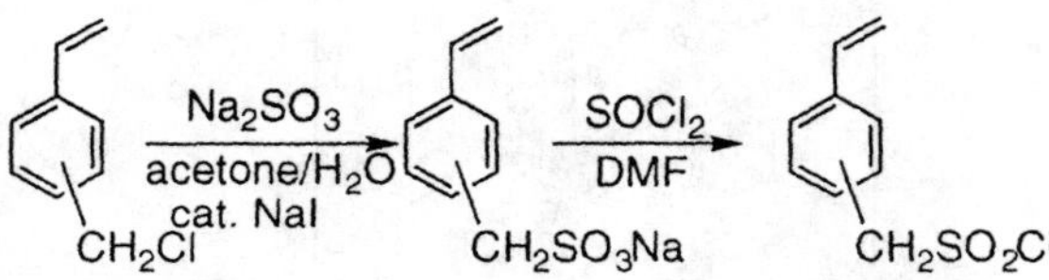
Na2SO3
acetone/H2O
cat. NaI
CH2Cl
SOCl2
DMF
CH2SO3Na
CH2SO2Cl

VBC (6) NaVBS (7) VBSC (8)

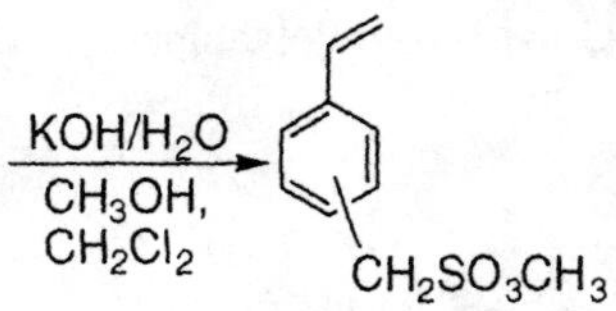
KOH/H2O
CH3OH,
CH2Cl2
CH2SO3CH3

MVBS (9)

Figure 4. Synthesis of MVBS.

employed as initiator and NaSS as charged monomer (A25N/25S), the precursor latex was stable, but during reaction with trimethylamine the latex coagulated.

Table I. Compositions of the Charged Stabilized Copolymers

Sample[a]	Weight of monomers, g			Stability	
	Styrene	VBC	SSC	Precursor	After $(CH_3)_3N$
C50N[b]	4.0	4.0	--	stable	stable
C25N/25S[b]	4.0	2.0	2.0	coagulated	unstable
C50N/50S[b]	--	4.0	4.0	coagulated	unstable
A25N/25S[c]	4.0	2.0	2.0	stable	unstable
A37.5N/12.5S[c]	3.2	3.2	1.6	stable	stable
A12.5N/37.5S[c]	3.2	1.6	3.2	stable	unstable

[a]C, A = cationic, anionic stabilized latexes; 25N/25S = wt % of monomer precursors to N^+ and SO_3^- repeat units.
[b]The sample contained 60 mg of VBTMAC, 200 mg of DVB and 70 mL of water at 60 °C and was initiated with 80 mg of VA-044.
[c]The sample contained 80 mg of NaSS, 100 mg of DVB and 65 mL of water at 50 °C, and was initiated with a solution of 70 mg of $K_2S_2O_8$, 27 mg of $NaHSO_3$ and 45 mg of $NaHCO_3$.

Only using excess VBC were stable precursor and polyampholyte latexes produced [A37.5N/12.5S, theoretical 3:1 (N^+/SO_3^-) charged latex, Table I], but the quaternized latex was polydisperse with particle diameters of 60-210 nm. Use of methyl m,p-vinylbenzyl poly(ethylene oxide) (MVBPEO, **8**, Figure 3) for steric stabilization of the terpolymers of styrene, VBC and SSC gave colloidally stable latexes (Table II).

Table II. Compositions of the Steric Stabilized Copolymer Latexes

Sample[a]	Styrene (g)	VBC (g)	SSC (g)	MVBPEO (g)
SS50N	3.6	4.0	-	0.4
SS25N/25S	3.6	2.0	2.0	0.4
SS50S	3.6	-	4.0	0.4

[a]SS = steric stabilized latexes. All samples contained 1 wt % of initiator VA-044, 0.10 g of DVB, 0.06 g of VBTMAC and 70 mL of water.

The control latex SS50N was iridescent during dialysis, indicative of colloidal crystallization of monodisperse particles, but the sulfonate containing latexes (SS25N/25S and SS50S) were never iridescent. TEM analyses revealed that SS50N was monodisperse with a diameter of 166 nm, and that SS25N/25S had diameters of 25-235 nm. Thus, the sulfonate containing latexes were polydisperse.

Since hydrolysis of SSC during polymerization might be responsible for colloidal instability and polydisperse latexes, its stability in water was tested as follows at 25 °C and at 60 °C (the temperature used for emulsion polymerization). D_2O was added to a solution of acetone/SSC to give a two-phase mixture, and the vinyl hydrogen signals were measured by ^{1}H NMR. The results in Figure 5 show > 90% conversion of the sulfonyl chloride to the *p*-styrenesulfonic acid in 75 min at 60 °C and 40 % conversion in 90 min at room temperature. At the end of the reaction the *p*-styrenesulfonic acid was completely soluble. Although these conditions were not identical to those of the emulsion polymerization, they do indicate that SSC monomer would at least partly hydrolyze during polymerization at 60 °C. The water soluble *p*-styrenesulfonic acid could form water soluble polymer and/or cause secondary particle nucleation resulting in polydisperse latexes with varied compositions.

Polymerizations of Methyl *p*-Styrenesulfonate (MSS, 3). A batch emulsifier-free polymerization of MSS, VBC and styrene (25:25:50 wt %) gave a viscous dispersion that would not pass through a cotton plug. Reaction with trimethylamine did not reduce the viscosity, which was probably due to water-soluble polymer.

Table III. Compositions of the semibatch copolymer latexes

Sample[a]	Styrene (g)	VBC (g)	MSS (g)	SSC (g)
SS25N	1.0	1.0	-	-
SS25N/25S	1.0	1.0	-	1.0
SS3N/1S	1.0	1.5	0.5	-

[a]SS = steric stabilized latexes. The initial reactor charge contained 1 wt % (40 mg) of initiator VA-044, 50 mL of water, 0.0076 g of VBTMAC, 0.051 g of MVBPEO, and 1.00 g of styrene. The monomer feed mixture also contained 0.15 g of MVBPEO and 0.05 g of DVB.

The failures of SSC and MSS to form colloidally stable monodisperse latexes by a batch method could be due to low solubilities of the sulfonyl monomers in the polymers, unfavorable copolymer reactivity ratios (*20,21*), or formation of water-soluble polymers. Semi-continuous addition of monomers, using a polystyrene seed and adding the SSC or MSS only during the seed growth stage, produced the compositions shown in Table III. The copolymers containing SSC and MSS coagulated after two weeks of storage.

Polymerizations of Methyl *m,p*-Vinylbenzylsulfonate (MVBS, 9). From concern that MSS might either have too high reactivity in copolymerization with styrene (*21*) or hydrolyze to styrenesulfonic acid during emulsion polymerization, we synthesized MVBS (**9**). As an alkylstyrene MVBS should have copolymerization reactivity similar to that of VBC, which forms approximately random copolymers with styrene (*22*).

Initial reactions of NaVBS with thionyl chloride (Figure 4) resulted in two vinylbenzyl monomers identified by their CH$_2$ signals in the ^{1}H-NMR spectrum. The by-product was identified as VBC, which must be produced by chloride ion displacement of a sulfonate leaving group from the benzylic carbon. The reactive sulfonate leaving group is probably formed from NaVBS and the Vilsmeier-Haack reagent $(CH_3)_2NCHCl^+Cl^-$ (*23*). VBC is produced thermodynamically and can be quenched without lost of yield by lowering the reaction temperature to -5 °C.

During storage of MVBS at 4 °C white crystals of the *para* isomer formed. The supernatant yellow oil enriched in the *meta* isomer was removed by washing with cold hexane, but recovery of each purified monomer was low. The NaVBS isomers were separated by washing the precipated solid with ethanol. A 85:15 *m/p* mixture of NaVBS was recovered from the ethanol solution. This mixture of sodium salts was treated with thionyl chloride to produce a 23% yield of 75/25 *m/p* MVBS.

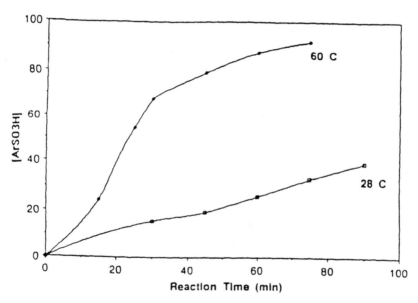

Figure 5. Hydrolysis of SSC.

The white solid left after the ethanol washing was >90% *para* NaVBS. Treatment of the *para* NaVBS with thionyl chloride produced a 44% yield of 17/83 *m/p* MVBS.

Semibatch emulsifier-free polymerization using the 50/50 *m/p* mixture of MVBS and a cationic initiator resulted in coagulated latexes. The 75/25 *m/p* MVBS was used for two different buffered one shot copolymerizations with styrene and VBC (Table IV). Using a mixture of NaSS and MVBPEO, the copolymer latexes before quaternization were stable, and no coagulum was collected after the polymerization. The latexes SS10N/10S containing 10 wt % each of VBC and MVBS and SS20N/20S containing 20 wt % each of VBC and MVBS, began to coagulate during dialysis after reaction with trimethylamine. TEM analysis of SS10N/10S revealed a bimodal particle size distribution with small 150 nm and larger 3 μm diameter particles. *para*-rich MVBS (83% para) was immiscible with styrene and VBC, and as a result, could not be copolymerized.

Table IV. Compositions of the MVBS copolymer latexes

Sample[a]	Styrene (g)	VBC (g)	MVBS (g)	MVBPEO (g)
SS10N/10S	1.5	0.2	0.2	0.1
SS20N/20S	1.1	0.4	0.4	0.1

[a]SS = steric stabilized latexes. The sample contained 15 mg of NaSS, 20 mg of DVB and 20 mL of water at 50 °C, and was initiated by a solution of 17.7 mg of $K_2S_2O_8$, 12 mg of $NaHSO_3$ and 17.8 mg of $NaHCO_3$.

Conclusions

None of the three styrenesulfonate monomers produced monodisperse colloidally stable latexes. SSC hydrolyzed competitively with emulsion polymerization, and the resulting *p*-styrenesulfonic acid caused secondary particle formation. MSS and *meta*-rich MVBS gave unstable and/or polydisperse latexes due to either low solubility in VBC/styrene latexes or too high reactivity during copolymerization. *para*-rich MVBS was incompletely miscible even with VBC and styrene monomers. Methacrylic acid is now under investigation as the source of negative charge in cross-linked polyampholyte latexes.

Acknowledgment. This research was supported by the U.S. Army Research Office.

Literature Cited

1. Higgs, P.G.; Joanny, J. *J.Chem. Phys.* **1991**, *94*, 1543.
2. Peiffer, D.G.; Lundberg, R.D. *Polymer* **1985**, *26*, 1059.
3. English, A.E.; Tanaka, T.; Edelman, E.R. *Macromolecules* **1998**, *31*, 1989.
4. McCormick, C.L.; Salazar, L.C.; Welch, P.M.; Mumick, P.S. *Macromolecules* **1994**, *27*, 323.
5. Tanaka, T.; Grosberg, A.Y.; Yu, A.; Manzanares, J.A.; Mafe, S. *J. Chem. Phys.* **1996**, *104*, 8713.
6. Candau, F.; Selb, J.; Corpart, J. *Polymer* **1993**, *34*, 3873.
7. Candau, F.; Neyret, S.; Candau, S.J.; Munch, J.P.; Skouri, M. *Macromolecules* **1994**, *27*, 69.
8. Candau, F.; Corpart, J. *Macromolecules* **1993**, *26*, 1333.
9. McCormick, C.L.; Kathmann, E.E.; *J. Polym. Sci: Part A* **1997**, *35*, 231.
10. McCormick, C.L.; Kathmann, E.E.; *J. Polym. Sci: Part A* **1997**, *35*, 243.
11. Ford, W.T.; Yu, H.; Lee, J.J.; El-Hamshary, H. *Langmuir* **1993**, *9*, 1698.

12. Ford, W.T.; *React. Funct. Polym.* **1997**, *33*, 147.
13. Kamogawa, H.; Kanzawa, A.; Kadoya, M.; Naito, T.; Nanasawa, M. *Bull. Chem. Soc. Jpn.* **1983**, *56*, 762.
14. Upson, D.A. *Macromolecular Syntheses* **1986**, *10*, 8.
15. Kim, J.H.; Chainey, M.D.; El-Aasser, M.S.; Vanderhoff, J.W. *J. Polym. Sci. Part A: Polym. Chem.* **1992**, *30*, 171.
16. Juang, M.S.; Krieger, I.M. *J. Polym. Sci. Polym. Chem. Ed.* **1976**, *14*, 2089.
17. Sunkara, H.; Jethmalani, J.M.; Ford, W.T. *J. Polym. Sci. Part A: Polym. Chem.* **1994**, *32*, 1431.
18. Takata, K.; Ihara, H.; Sata, T. *Angew. Makromol. Chem.* **1996**, *236*, 67.
19. Shahar, M.; Meshuham, H.; Margel, S. *J. Polym. Sci. Polym. Chem. Ed.* **1986**, *24*, 203.
20. Jerabek, K.; Hankova, L. *Ind. Eng. Chem. Res.* **1995**, *34*, 2598.
21. Woeste, G.; Meyer, W.H.; Wegner, G. *Makromol. Chem.* **1993**, *194*, 1237.
22. Brandrup, J.; Immergut, E.H., Eds., *Polymer Handbook*, 3rd ed.; Wiley: New York, 1989; p II-225.
23. Busshard, H.H.; Mory, A.; Schmid, M.; Zollinger, H. *Helv. Chim. Acta*, **1959**, *42*, 1653.

Chapter 4

Water Soluble Polymers Produced by Homogeneous Dispersion Polymerization

Patrick G. Murray and Manian Ramesh

Nalco Chemical Company, Global Polymer Science,
One Nalco Center, Naperville, IL 60563–1198

The preparation of high molecular weight, water soluble acrylamide-based flocculants as water continuous dispersions is described. This innovative method of manufacture eliminates many of the undesirable characteristics associated with the production and application of these flocculants as conventional water-in-oil emulsions or as dry powders. The monomers and their polymers, the role of the stabilizer polymer, particle characteristics, viscosity considerations, and the thermodynamic and physical stability of these polymer dispersions is discussed.

Water is a primary raw material or processing aid in most industrial operations, including paper manufacturing, oil production, municipal waste treatment, mining and mineral processing, petroleum refining, metalworking, steam generation, and food and beverage production. Because of its unique physical properties and relative abundance, water is used in all of these industries to transport solids economically or to transfer heat efficiently. Typically, water taken from the environment for use in these operations needs to be conditioned to remove suspended solids, organic matter and other materials that may be detrimental to the process. Likewise, water that is being used in these industrial processes must often be treated at some point to effect a solid/liquid or a liquid/liquid separation, which is oftentimes critical to the efficiency of the central operation. Finally, water that has been so used must often once again be treated prior to its return to the environment in order to remove harmful wastes and contaminants (oils, sludges, metals, etc.). Indeed, various regulations that place strict requirements on the quality of wastewater discharges are now in place in most industrialized nations. Synthetic organic polymers are widely used for all of the above listed purposes. A substantial fraction of these polymers, which are now consumed at the rate of several hundred million pounds per year, with a market value of approximately one billion dollars, are produced as high molecular weight copolymers of acrylamide (1, 2).

Until recently, the manufacture of high molecular weight acrylamide polymers or copolymers was accomplished by one of two methods. The polymers could be produced as water-in-oil emulsions (or "latex" polymers), consisting of a hydrocarbon-based solvent for the continuous phase and various surfactants to provide emulsion stability. Polymers prepared in this fashion are generally 20% to

40% active polymer. As liquid products, the water-in-oil emulsions are easy to handle and relatively easy to prepare for their end use, given the proper equipment. Despite these advantages, the water-in-oil emulsions do have several inherent disadvantages. For example, the hydrocarbon continuous phase and surfactant systems that enable their manufacture as liquids play no role in the end use application of the active polymer. As a consequence, at the current consumption levels of about 80 million pounds of active polymer per year, about 90 million pounds of these oils and surfactants are introduced into the environment along with the active polymer, since, until now, no improvements in the manufacture of these polymers as liquids have been forthcoming. Moreover, at the end use application, these water-in-oil emulsions are "inverted," or diluted with water and the emulsion destabilized by the addition of yet an another surfactant in order to prepare the polymer component for use. The second primary method for the manufacture of high molecular weight acrylamide-based polymers has been to make solutions or gels which are subsequently dried and ground to a fine powder. While no hydrocarbon or surfactant is employed in the manufacture of polymers in this form, the drying step is energy intensive, and dissolution of the dry powder at the end use application is often tedious. In addition, because the polymers cannot be pumped from closed storage containers like the water-in-oil emulsions, care must be taken in their handling in order to limit exposure of plant personnel to dust.

In order to overcome the difficulties associated with inverse emulsion and dry polymers, Nalco has become involved in the development and commercial practice of a unique technology for the manufacture of high molecular weight water soluble polymers based on acrylamide. This polymerization process permits the manufacture of these extremely useful polymers as water continuous dispersions. The polymer products are liquid, and so retain the virtues of ease and safety of handling, but they are manufactured in water instead of in a hydrocarbon and surfactant matrix. Thus, no oil or surfactants are released to the environment with the application of these polymers. The performance of these polymers in the various end use applications is equivalent to, or in some cases exceeds, that obtained with similar polymers produced in inverse emulsion or dry form. A discussion of this dispersion polymerization technology, the monomers and their polymers, the stabilizer polymers, particle characteristics, viscosity considerations and the thermodynamic and physical stability of the products constitutes the subject of this manuscript.

Background

While a comprehensive review of conventional, heterogeneous dispersion polymerization is beyond the scope of this manuscript, a brief recital of fundamental points will facilitate the discussion to follow. Dispersion polymerization in any form is essentially a precipitation polymerization, though distinguished from this broad characterization in the sense that discrete, well defined polymer particles are produced. One unique feature of dispersion polymerization is that the monomers, a stabilizer polymer and any other components of the final product dissolve completely in what is eventually to become the continuous phase. The polymer that forms, however, is insoluble in this solution, and as a consequence precipitates. The second distinguishing feature of a dispersion polymerization involves the presence of the stabilizer polymer. This preformed polymer, present from the beginning of the dispersion polymerization, allows for the production of uniform polymer particles by preventing agglomeration both during and after the polymerization. It is generally believed that the stabilizer polymer either adsorbs to or, some cases, may become grafted (3) to the polymer particles, and in so doing provides a mechanism for steric or electrostatic stabilization of the dispersion. Since the components of the formulation are all initially soluble, the polymerization begins in solution, though eventually the polymer phase becomes discontinuous and the polymerization proceeds to completion as a heterogeneous system. Not surprisingly, there has been a great deal of interest in heterogeneous dispersion polymerization since its inception, as evidenced by the volume of academic and patent literature now to be found relating to the subject (4-7).

In a classical heterogeneous dispersion polymerization, the continuous phase is organic in nature, although in some instances water has been used as a component of the continuous phase to increase polarity. Research investigations have focused on the composition of the dispersion medium, reaction kinetics, the structure and influence of the stabilizer polymer, particle size, molecular weight and molecular weight distribution (4, 5). Dispersions of poly(methyl methacrylate) (8) and poly(styrene) (9) are widely studied and among the best characterized systems. Recently, dispersion polymerizations conducted in supercritical carbon dioxide have also been reported (9-12).

Homogeneous Dispersion Polymerization

Several features distinguish the preparation of water soluble polymers dispersed in an aqueous continuous phase from the more conventional heterogeneous dispersion polymerization discussed above. For example, the dilution of a hexane continuous poly(methyl methacrylate) dispersion with more hexane will simply dilute the dispersion; the addition of more water (or salt water) to a water continuous dispersion of water soluble polymer will completely destroy the dispersion stability and result in an extremely high viscosity polymer solution. Indeed, this property is the basis for the ready application of these polymers as water treatment flocculants. In the case of the typical heterogeneous dispersion, poly(methyl methacrylate) in hexane, for example, it is known that very little high molecular weight polymer exists outside of the particles. In contrast, the amount of polymer present in the continuous phase of a water continuous dispersion of water soluble polymer is entirely a function of the ionic strength of the continuous phase. This must be controlled, of course, and impacts in-process and post-polymerization (product) viscosities dramatically. In short, the formation of the water continuous dispersion of water soluble polymers is reversible. Thus, in order to recognize these distinguishing features of the technology, we have elected to call this process of manufacturing water continuous dispersions of water soluble polymers *homogeneous* dispersion polymerization.

The first examples of homogeneous dispersion polymerization were reported by Takeda, et al., (13, 14) although manufacturing improvements (15-19) and some academic investigations (20, 21) were soon forthcoming from others. While we will avoid the reproduction of specific recipes herein, the experimentalist may derive such information from the work of Takeda. From a fundamental perspective, however, it is probably most useful to begin a broad based discussion with the three essential components of any homogeneous dispersion: the aqueous dispersing medium, the water soluble monomers and the water soluble stabilizer polymer.

The Continuous Phase. As has already been described, the continuous phase of any dispersion polymerization must be a good solvent for the monomers, the stabilizer polymer and the initiator, but a poor solvent for the polymer being prepared. For the preparation of homogeneous dispersion polymers, this is accomplished by adjusting the ionic strength of the water used for the continuous phase by the addition of one or more salts. In principle, any water soluble salt might be employed for this purpose, including chlorides, phosphates, sulfates, nitrates, and so on. In our own laboratories, the sulfates have generally been the most useful. Salts of organic acids are also potentially useful, provided that the pK_a of the acid is sufficient to solubilize it at the pH required by the polymerization and that no other functionality in the acid interferes with the subsequent free radical polymerization. Judicious selection of the salt type and concentration is undoubtedly a primary requirement for producing a stable homogeneous dispersion, and oftentimes this information must be derived by experiment if no analogous systems exist. Most often, the salt concentration that is optimum during the polymerization is different from the salt concentration that will result in a physically stable dispersion over long periods of time. This aspect of these systems will be discussed in detail later. Finally, the pH of the continuous phase must be compatible with the monomers to be polymerized. Homogeneous dispersion polymerizations may be initiated using either redox or any of several water soluble azo initiators.

Some viscosity may be expected to develop during the "solution polymer" stage of the polymerization. While the extent and duration of the viscosity that can be tolerated is a function of the agitation efficiency and heat removal capability of the reactor, the salt concentration is the primary parameter to be adjusted in order to obtain the optimum in-process viscosity. If the salt concentration has been properly adjusted, then at the onset of the precipitation event the viscosity of the system drops markedly and the polymerization is finished as a dispersion. If the salt concentration is too low during the polymerization, the precipitation event and transition to a heterogeneous system never occurs, and effective mixing and heat transfer will eventually be lost. If, on the other hand, the salt concentration is too high, the dispersion will have little thermodynamic stability in the absence of mechanical shear.

The Monomers and Their Polymers. Cationic copolymers of acrylamide are widely used as water treatment agents, being prepared in both inverse (water-in-oil) emulsions or solution polymers that are dried and ground. These compositions are generally derived by polymerizing acrylamide with quaternary amine containing acrylic esters or amides, the methyl chloride quaternary salt of dimethylaminoethyl acrylate (DMAEA.MCQ) being perhaps the most ubiquitous comonomer. The structure of this monomer, DMAEA.MCQ, is shown in Figure 1.

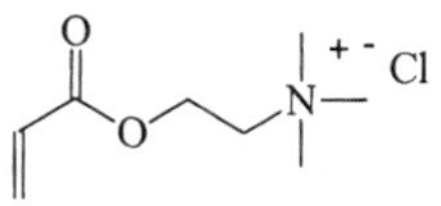

Figure 1. Dimethylaminoethylacrylate, methyl chloride quaternary salt

In the preparation of homogeneous dispersions, where the objective is to obtain a precipitated polymer from water soluble monomers, it is useful to consider monomers that are water soluble, but somewhat more hydrophobic than DMAEA.MCQ and therefore more likely to render the polymer insoluble in the high ionic strength continuous phase. One monomer that facilitates the precipitation of the polymer because of its increased relative hydrophobicity is the benzyl chloride quaternary salt of dimethylaminoethyl acrylate, or DMAEA.BCQ, the structure of which is shown in Figure 2.

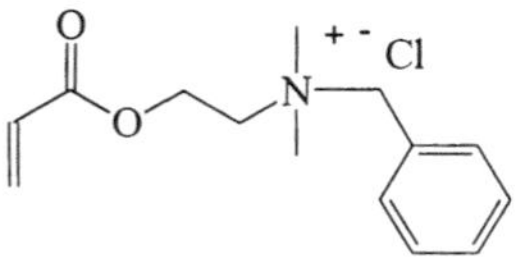

Figure 2. Dimethylaminoethylacrylate, benzyl chloride quaternary salt

By ensuring that some or all of the cationic charge of these acrylamide copolymers is derived from DMAEA.BCQ, instead of DMAEA.MCQ, a fluid, low viscosity dispersion can be prepared with the same overall amount of cationic charge as those polymers prepared as inverse emulsions or as dry powders. The reader will recall that it is necessary for the monomers to dissolve fully in the salt solution prior to polymerization, and that, in order for the polymer to be useful as a flocculant, it must eventually be able to be re-dissolved completely. For these reasons, hydrophobic, water insoluble monomers like vinyl acetate or styrene, for example, have generally not been useful in this capacity. However, as an understanding of the factors that influence viscosity during these homogeneous dispersions has been obtained, it has become possible to manufacture dispersions without DMAEA.BCQ monomer (22). This is advantageous, as it permits the polymers to be used in certain FDA regulated markets, like papermaking, for example (23).

Homogeneous dispersions with polymer concentrations between 15% and 30% are typical. At less than 15% polymer solids, the manufacturing and shipping costs of the products become unattractive; at greater than 30% polymer solids, the in-process polymerization viscosities or final product viscosities can become too high for routine handling.

The Role of the Stabilizer Polymer. The composition and structure of the stabilizer polymer profoundly influence the course of a homogeneous dispersion polymerization. This polymer must stabilize incipient and growing polymer particles against aggregation during the polymerization, and must provide a thermodynamic resistance to coalescence of mature particles after the polymerization is complete. While structure-activity relationships have been extensively studied for heterogeneous dispersion systems, relatively little fundamental mechanistic work has been conducted with homogenous dispersion polymerizations. The primary reason for this is that most of the research on these systems to date has been industrial, and not academic. As a consequence, the polymers selected for study have, for the most part, been limited to economically viable and commercially accessible alternatives. Thus, the stabilizer systems in use today to manufacture homogeneous dispersions in large-scale reactors are homopolymers or copolymers of widely used and readily available water soluble monomers like the quaternary ammonium salts of acrylate esters, diallyldimethylammonium chloride (DADMAC), acrylamide, acrylic acid, or acrylamidomethylpropane sulfonate (AMPS) and related monomers. In general, stabilizers with net cationic charge are used to prepare homogeneous dispersions of cationic copolymers of acrylamide, and stabilizers with net anionic charge are used to prepare dispersions of acrylamide and anionic monomers.

One of the most important advances in the production of homogeneous dispersion polymers was the development of less hydrophilic stabilizers (17). For example, copolymers of hydrophilic monomers like diallyldimethylammonium chloride and water soluble, more hydrophobic monomers like DMAEA.BCQ have been prepared and used to manufacture a variety of cationic copolymers of acrylamide under low viscosity conditions. It is thought that the more hydrophobic stabilizer systems function by decreasing the hydrodynamic volume of the polymer particles in solution, thereby lowering the viscosity observed during the polymerization. As will be discussed shortly, the control of in-process viscosity is central to the successful manufacture of homogeneous dispersions, particularly in commercial equipment. As such, a stabilizer that lowers the viscosity of the system during the polymerization allows for more effective mixing and better heat removal, which can translate to larger batch sizes, faster reactions (shorter cycle times) and better molecular weight control.

As has been reported elsewhere, block and graft copolymers have been shown to be excellent stabilizers for many heterogeneous dispersions. Therefore, it is not surprising that an interest in block and graft copolymers as stabilizers for certain homogeneous dispersion polymerizations has also developed (25, 26). These investigations have been mainly directed toward exploring the effect of stabilizer structure on particle size and reaction kinetics (solution versus dispersed phase polymerization rates).

The concentration of the stabilizer polymer in homogeneous dispersions typically ranges from about 3% to 6% by weight based on monomer. The precise amount of stabilizer required is generally determined empirically, and can be optimized to provide desirable properties such as low gel (coagulum) and uniform particle size to the final product. The concentration of the stabilizer polymer in the dispersion formula can also impact the final viscosity of the product to the extent that the stabilizer (or some portion of it) is dissolved in the continuous phase.

In-Process and Final Product Viscosity Characteristics. As has already been observed, the formation of a homogeneous dispersion is reversible. That is, by the addition of more water to the continuous phase, the dispersion can be made to revert to a high viscosity solution polymer, and by the addition of more salt to this high viscosity solution it can be made to go back into a dispersion. This unique feature of the homogeneous dispersion must be carefully controlled, as excessively high in-process viscosity during polymerization is generally not a desirable feature, particularly in commercial-scale equipment. The maximum viscosity that is observed during the polymerization will dictate, to a large extent, what equipment is suitable for routine manufacture. However, in order to use reactors equipped with conventional (pitched-blade) axial-flow impellers, the viscosity should be kept below 10,000 centipoise. The salt concentration in the continuous phase and the choice of the polymeric stabilizer are the primary variables that can be used to meet this criterion.

Although the viscosities observed during these polymerizations can be quite high (impeller viscosities of 10,000 centipoise or greater), the viscosities of the final products are usually 100-1000 centipoise. Moreover, until high conversion is reached, the homogeneous dispersions often exhibit non-Newtonian, shear thinning fluid behavior. In these instances, indirect measures of fluid viscosity such as the power draw of the agitator motor or the efficiency of heat transfer tend to become useful, since the power input characteristics and mixing efficiencies of various impellers change considerably with fluid viscosity. A typical viscosity profile for a homogeneous dispersion polymerization is shown qualitatively in Figure 3.

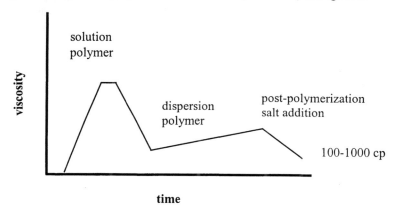

Figure 3. **Typical viscosity profile for homogeneous dispersion polymerization**

As can be seen in Figure 3, an increase in the viscosity of the system is typically observed after initiation, as high molecular weight solution polymer is formed. The viscosity of the system then drops as a dispersion is formed, and the conversion to polymer is finished as a much lower viscosity heterogeneous system. It is not unusual for the viscosity of these polymerizations to increase slightly during this period, although, with the addition of more salt after the polymerization, the final product will oftentimes have a viscosity of only a few hundred centipoise. This post-polymerization salt addition is discussed in further detail below, under Dispersion Stability.

As would be expected, polymers made with the relatively hydrophobic DMAEA.BCQ monomer are less soluble in 1M sodium nitrate solution than an analogous polymer wherein the cationic charge is derived from DMAEA.MCQ. For these reasons, Reduced Specific Viscosity (RSV) and Intrinsic Viscosity (IV) measurements for these polymer are conducted in 0.125M sodium nitrate solution. RSVs of dispersions made without DMAEA.BCQ monomer can be measured in the same solvents and at the same concentrations used for the inverse emulsion polymers of the same composition. Under these conditions, the RSV and IV values obtained for polymers made by homogeneous dispersion polymerization are comparable to their latex counterparts. Absolute molecular weight information can be obtained using Multi-Angle Laser Light Scattering (MALLS); the molecular weights of the homogeneous dispersion polymers listed in Table I have analyzed as being within ± 5% of their latex counterparts in our laboratories using this technique.

The Polymer Particles. The polymer particles produced by homogeneous dispersion polymerization are usually on the order of several microns in size, and particle size distributions can be unimodal or bimodal, depending upon the polymerization conditions. Polymer particles smaller than about one micron tend to agglomerate to larger, amorphous clusters, which can result in dispersion instability. On the other hand, the viscosity of the product is strongly dependent upon the particle size, and so there is a practical limitation on how large the particles should be for optimum handling characteristics. Besides the stabilizer concentration, the salt concentration and the agitation power (mechanical shear) also influence the particle size that is obtained, and all of these parameters must be optimized for long-term dispersion stability. A comparison of some average particle sizes for a series of homogeneous dispersions is presented in Table I. Particle size data for dispersion polymers were obtained by optical microscopy and/or laser diffraction; data for inverse emulsions were collected using a centrifugal particle size analyzer.

Table I. Particle size comparisons between acrylamide-based polymers prepared by inverse emulsion and homogeneous dispersion polymerization

Polymer	Composition	d50 (μm)
inverse emulsion	90/10 mole % acrylamide/DMAEA.MCQ	1-2
dispersion	90/10 mole % acrylamide/DMAEA.MCQ	4-5
dispersion	90/10 mole % acrylamide/DMAEA.BCQ	9-14
inverse emulsion	50/50 wt % DADMAC/acrylamide	1-2
dispersion	50/50 wt % DADMAC/acrylamide	3-4

Dispersion Stability. Given that the proper stabilizer polymer and optimal salt concentration have been selected, polymer particles produced by homogeneous dispersion can resist flocculation indefinitely. However, a second consideration related to the stability of the dispersion is the tendency of the particles (which are "dispersed" in the continuous phase) to either rise or settle. It is the relative rates of polymer settling or floating in the continuous phase that are of concern when the dispersion is expected to maintain a uniform polymer concentration over time throughout an unagitated storage vessel, for example.

The velocity (rate) of particle sedimentation in a fluid medium is given by Stokes' law,

$$v = 2a^2(\rho_2 - \rho_1)g / 9\eta$$

where a is the diameter of the particle, η is the viscosity of the fluid medium, g is the gravitational force, and ρ_1 and ρ_2 are the densities of the fluid medium and the particle, respectively. The application of Stokes' law to the water continuous dispersion polymers is an oversimplification, since, among other things, it neglects particle-particle interaction. Nonetheless, it is useful, as it defines the parameters that may be exploited in order to alter the rate of polymer particle settling or floating. As is evident from Stokes' equation, the size of the particle, the difference between the densities of the dispersed and continuous phases, and the viscosity of the continuous phase all contribute to the rate of particle migration in the fluid medium. In order to slow the rate of migration, then, one can 1) make smaller particles, 2) increase the viscosity of the fluid medium (continuous phase), or 3) minimize the difference in the densities of the two phases.

In the case of water continuous dispersion polymers, the most easily manipulated of these three parameters is the third: the density of the continuous phase can be adjusted so as to match the density of the polymer phase. Generally, an increase in the density of the continuous phase is required, which can be accomplished by the addition of more salt to the dispersion after the polymerization is complete. Typically, the salt used for post-polymerization adjustment of the continuous phase density is the same as, or similar to, the salt used to effect precipitation of the polymer particles as they are formed during the polymerization. If the other parameters (particle size and viscosity of the continuous phase) are not altered dramatically by slight adjustments in the density of the continuous phase, and the density of the particles does not change over time, then a product in which the densities of the polymer phase and the continuous phase are matched should exhibit commercially useful settling stability.

In order to determine the optimum final salt concentration for long-term settling stability, the homogeneous dispersions can be subjected to high speed centrifugation to intentionally effect a separation of the polymer particles from the continuous phase. A gravitational force of 8000g for several minutes is sufficient to accomplish the separation in most cases. If the densities of the polymer phase and the continuous phase are perfectly matched, then no separation occurs. However, if the polymer phase is more dense than the continuous phase, the polymer will settle to the bottom of the centrifuge tube (sediment) and the liquid continuous phase can be decanted. Thus, a method to optimize the settling properties of the dispersion can be envisioned wherein the salt concentration is incrementally increased until no separation of polymer from the continuous phase in the centrifuge is observed. In our laboratories, this method is routinely used to optimize long-term settling stability of the dispersions during process development.

Safety Considerations. As with any polymerization, safety concerns related to the preparation of water soluble polymers derived from acrylamide and related monomers generally fall into two categories: those associated with toxicity of the monomers, and those related to the exothermicity of polymerizing the monomers. Regarding the former, all precautions associated with the safe handling of acrylamide monomer should be followed. Acrylamide monomer is toxic, and so all polymers should be treated as containing residual monomer until analysis is available which indicates high conversion.

The exothermicity of a given polymerization is, of course, a function of a number of things including the heats of polymerization of the monomers, the amount of monomer in the polymerization formula, and the heat capacities of the other ingredients of the polymerization formula. The primary concern is that of a runaway reaction, when temperature control is lost and the capability to exceed the boiling point of one or more of the components of the formulation is present. It is certainly possible to assemble formulations of water soluble monomers that contain enough latent heat to achieve this condition, but the preparation of polymers under homogeneous dispersion conditions presents an additional complication. In the event of a runaway homogeneous dispersion polymerization, it is likely that agitation in the reactor will be lost at some point, and that the contents of the reactor will solidify. If this happens with a mixture of monomers with enough latent heat to reach the boiling point of water, steam becomes trapped within a solid mass, and reactor venting assumptions based on a low viscosity inverse emulsion polymer of the same monomer composition are probably optimistic. Under these circumstances, a containment strategy using a reactor that is rated for pressure is recommended.

Acknowledgments

The authors wish to thank Cindy McCullagh of Nalco for her contribution of the particle size data in Table 1 and collection of the molecular weight information using MALLS.

Literature Cited

(1) Rosen, M. R.; *CMR Focus Report*, October 13, 1997, 13.
(2) Peaff, G.; Chem. Eng. News, November 14, 1994, 15.
(3) Grafting of the stabilizer polymer has been observed with certain cellulosic polymers (see Paine, A. J. *J. Colloid Interface Sci.*, **1990**, 138, 157, for example) or with macromonomers used as steric stabilizers (see Kawaguchi, S.; Winnik, M. A.; Ito, K. *Macromolecules*, **1995**, 28, 1159, for example).
(4) Sudol, E. D. in *Polymeric Dispersions: Priciples and Applications*; Asua, J. M., Ed.; Kluwer Academic: Boston, MA, 1997; p. 141.
(5) Croucher, M. D.; Winnik, M. A., in *Scientific Methods for the Study of Polymer Colloids and Their Applications*, Candau, F. and Otteweill, R. H., Eds.; Kluwer Academic Publishers: Amsterdam, 1990, p. 35.
(6) Barrett, K. E. J.; *Dispersion Polymerization in Organic Media*; Wiley: London, 1975
(7) Napper, D. H.; *Polymeric Stabilization of Colloidal Dispersions*; Academic Press: London, 1975
(8) Winnik, M. A.; Lukas, R.; Chen, W. F.; Furlong, P.; Croucher, M. D. *Makromol. Chem., Macromol. Symp.*, **1987**, 10-11, 483.
(9) see for example, Canelas, D. A.; Desimone, J. M. *Macromolecules* **1997**, 30, 5673, and references therein.
(10) Desimone, J. M.; Maury, E. E.; Menceloglu, Y. Z.; McClain, J. B.; Romack, T. R.; Combes, J. R. *Science* **1994**, 257, 945.
(11) Desimone, J. M.; Maury, E. E.; Menceloglu, Y. Z. U.S. Patent 5,382,623, 1995.
(12) O'Neill, M. L.; Yates, M. Z.; Johnston, K. P.; Smith, C. D.; Wilkinson, S. P. *Macromolecules*, **1998**, 31, 2838.

(13) Takeda, H.; Kawano, C. U.S. Patent 4,929,655, 1990.

(14) Takeda, H.; Kawano, C. U.S. Patent 5,006,590, 1991.

(15) Hurlock, J. R.; Ramesh, M. U.S. Patent 5,597,859, 1997.

(16) Werges, D. L., Ramesh, M. Eur. Patent 657,478, 1996.

(17) Ramesh, M., Cramm, J. R., Werges, D. L., Howland, C. P. U.S. Patent 5,597,858, 1997.

(18) Ramesh, M., Howland, C. P., Cramm, J. R. Eur. Patent 630,909, 1994.

(19) Selvarajan, R.; Hurlock, J. R. WO patent 9,734,933, 1997.

(20) Zimmerman, A.; Jaeger, W.; Reichert, K. H. *Polym. News*, **1997**, 22, 390.

(21) Jaeger, W.; Hahn, M.; Lieske, A.; Zimmerman, A., Brand, F. *Macromol. Symp.* **1996**, 111, 95.

(22) Wong Shing, J. B.; Tubergen, Karen R. U. S. Patent 5,750,034, 1998.

(23) Wong Shing, J. B.; Hurlock, J. R. Eur. Patent A2 831,177, 1998.

(24) Wong Shing, J. B. Eur. Patent A1 821,099, 1998.

(25) Hildebrandt, Volker; Reichert, Karl-Heinz, *Angewandte Makromolekulare Chemie*, **1997**, 245, 165.

(26) Jaeger, W.; Zimmerman, A.; Reichart, K-H.; Zeitz, K. Ger. Patent 1,9521,096, 1996.

Chapter 5

Emulsion Polymerizations with 2-Acrylamido-2-methylpropanesulfonic Acid

Geoffrey P. Marks[1] and Alan C. Clark[2]

[1]Lubrizol International Laboratories, P.O. Box 88,
Belper, Derby DE56 1QN, United Kingdom
[2]The Lubrizol Corporation, 29400 Lakeland Boulevard,
Wickliffe, OH 44092–2298

There is a growing need to improve the properties of latices to be used in applications such as paper coatings, pressure-sensitive adhesives, paints, carpet backing and personal care formulations. Desired attributes include improved compatibility of latices with divalent cation salts such as calcium carbonate, better latex particle mechanical stability under high shear conditions to avoid coagulation, improved adhesion and enhanced durability of coatings towards water washing. Three types of latex formulations have been prepared with sodium 2-acrylamido-2-methyl-1-propanesulfonate, AMPS® sodium salt monomer[1], replacing most of the surfactants and acrylic acid in an acrylic latex, replacing sodium vinyl sulfonate in a vinyl acrylic latex and replacing methacrylic acid in a styrene acrylic latex. The divalent cation stability of both the acrylic and the styrene acrylic latices were significantly improved by replacing the carboxylate monomers with AMPS sodium salt. The scrub resistance of paints made from the acrylic and styrene acrylic latices prepared with AMPS sodium salt were also significantly improved.

Three types of latex formulations have been prepared. The first formulation is an acrylic latex from methyl methacrylate, butyl acrylate and acrylic acid using sodium lauryl sulfate and Synperonic® NP20[2] (an ethoxylated nonylphenol) as surfactants. The second formulation is a vinyl acrylic latex from vinyl acetate, butyl acrylate and sodium vinyl sulfonate with sodium lauryl sulfate as the surfactant. The third formulation is a styrene acrylic latex from styrene, butyl acrylate and methacrylic acid with sodium lauryl sulfate as the surfactant. Three experimental latex formulations analogous to the three above were also prepared with AMPS sodium salt replacing acrylic acid, sodium vinyl sulfonate and methacrylic acid respectively.

Experimental

Each of the following six latex preparations was prepared by a semi-continuous, two-step procedure. In step one, 5-10%w of the monomers were added to water containing all of the required surfactant and more than 5-10w% of the catalyst. In step two, additional catalyst was added and the remainder of the monomers was added continuously over several hours. When AMPS sodium salt was one of the monomers, it was added continuously as a separate feed from the other monomers.

Baseline Acrylic Latex Preparation. A solution of sodium lauryl sulfate(3.53 g), Synperonic NP20 (10.36 g) and sodium bicarbonate (1.43 g) in 420 grams of water was prepared and purged subsurface with nitrogen to remove oxygen. The solution was heated to 80°C and a mixture of methyl methacrylate (37.8 g, 0.378 mol), butyl acrylate (30.8 g, 0.24 moles) and acrylic acid (1.4 g, 19 mmol) was added. After the temperature returned to 80°C, a solution of sodium persulfate (0.7g, 2.94 mmol) in 10 g of water was added. After 30 minutes, A mixture of methyl methacrylate (340 g, 3.40 moles), butyl acrylate (277 g, 2.16 moles) and acrylic acid (12.6 g, 0.17 moles) was added at a rate of 280 g per hour. After 30 minutes of monomer addition, a solution of sodium persulfate (3.5 g, 14.7 mmol) in 140 g of water was added at a rate of 55 grams per hour. When addition of all of the ingredients was complete, the latex was stirred under nitrogen for an additional 30 minutes, decanted and cooled.

Experimental Acrylic Latex Preparation. Prepare a solution of sodium lauryl sulfate (0.36 g) and sodium bicarbonate (1.4 g) in 403 grams of water at 80°C and purge the solution subsurface with nitrogen to remove oxygen. Add a mixture of methyl methacrylate (38.0 g, 0.38 moles) and butyl acrylate (31.2 g, 0.243 moles) to the solution. Add aqueous 50% AMPS sodium salt (1.4 g, 3.06 mmol) to the solution. After the temperature returned to 80°C, a solution of sodium persulfate (0.7, 2.94 mmol) in 10 g of water was added. After 30 minutes, a mixture of methyl methacrylate (342 g, 3.42 moles) and butyl acrylate (281 g, 2.19 moles) was added at a rate of 280 grams per hr. Aqueous 50% AMPS sodium salt (12.6 g, 27.5 mmol) in 140 g of water was added simultaneously at 70 g per hour. After 30 minutes of monomer addition, a solution of sodium persulfate (3.5 g, 14.7 mmol) in 140 g of water was added at a rate of 55 grams per hour. When addition of all of the ingredients was complete, the latex was stirred under nitrogen for an additional 30 minutes, decanted and cooled. Table 1 compares the baseline and experimental acrylic latex formulations.

Baseline Vinyl Acrylic Latex Preparation. A solution of sodium lauryl sulfate (9.1 g) and sodium bicarbonate (1.34 g) in 428 grams of water was prepared and heated to 84°C. Purge the solution subsurface with nitrogen to remove oxygen. After one hour of nitrogen purging, a mixture of vinyl acetate (18.0 g, 0.21 moles) and butyl acrylate (4.51 g, 0.035 moles), potassium persulfate (2.04 g, 7.55 mmol) and sodium vinyl sulfonate monomer (0.90 g, 1.73 mmol) was added. SVS was supplied as a 25% aqueous solution. After the reflux subsided, a mixture of vinyl acetate (343 g, 3.98 moles) and butyl acrylate (85.7 g, 0.665 moles) was added to the reaction mixture at

120 g per hour simultaneously with a mixture of SVS (17.1 g, 32.9 mmol) with water (40.9 g). The reaction mixture was stirred for an additional 30 minutes. A mixture of sodium formaldehyde bisulfite (0.18 g, 0.67 mmol) with water (1.3 g) was added in one-third portions every 10 minutes. A solution of *t*-butylhydroperoxide (0.18 g, 1.40 mmol) in water (1.3 g) was added in one-half portions at the same time as the first two sodium formaldehyde bisulfite additions. Once the additions were complete the temperature was lowered, the nitrogen purge was removed, and the mixture was allowed to cool to room temperature overnight.

Experimental Vinyl Acrylic Latex Preparation. A solution of sodium lauryl sulfate (9.1 g) and sodium bicarbonate (1.34 g) in 428 grams of water was prepared. Purge the solution subsurface with nitrogen to remove oxygen and heat to 84°C. After one hour of nitrogen purging, a mixture of vinyl acetate (18.0 g, 0.21 moles) and butyl acrylate (4.51 g, 0.035 moles), potassium persulfate (2.04 g, 7.55 mmol) and 50% aqueous AMPS sodium salt (0.45g, 0.98 mmol) was added. After the reflux subsided, A mixture of vinyl acetate (343 g, 3.98 moles) and butyl acrylate (85.7 g, 0.665 moles) was added to the reaction mixture at 120 g per hour simultaneously with 50% aqueous AMPS sodium salt (8.55g, 18.7 mmol) in 40.9 g water. The reaction mixture was stirred for an additional 30 minutes. A mixture of sodium formaldehyde bisulfite (0.18 g, 0.67 mmol) with water (1.3 g) was added in one-third portions every 10 minutes. A solution of *t*-butyl -hydroperoxide (0.18 g, 1.40 mmol) in water (1.3 g) was added in one-half portions at the same time as the first two sodium formaldehyde bisulfite additions. Once the additions were complete, the temperature was lowered, the nitrogen purging was stopped and the mixture allowed to cool to room temperature overnight. Table 2 compares the baseline and experimental vinyl acrylic formulations.

Baseline Styrene Acrylic Latex Preparation. A solution of sodium lauryl sulfate (0.364 g) and sodium bicarbonate (1.4 g) in 395 g of water was prepared, adjusted to pH=9 by the addition of 15 drops of a 20%w NaOH solution and heated to 80°C. Purge the solution subsurface with nitrogen to remove oxygen. Note: If the pH adjustment is not made, the product latex is coagulated. A mixture of styrene (36.0 g, 0.346 moles) and butyl acrylate (32.6 g, 0.254 moles) was added to the solution. A solution of methacrylic acid (0.7g, 8.13 mmol) neutralized with 20% NaOH in 14.6 g of water was added to the solution. When the temperature returned to 80°C, a solution of sodium persulfate (0.70 g, 2.94 mmol) in 10 grams of water was added. After 30 minutes, a mixture of styrene (324 g, 3.11 moles) and butyl acrylate (293.4 g, 2.286 moles) was added at a rate of 279 grams per hour. A solution of methacrylic acid (6.3g, 73.2 mmol) neutralized with 20% NaOH in 131.7 g of water was added to the solution as a separate feed. After 30 minutes of monomer feed, a solution of sodium persulfate (3.5 g, 14.7 mmol) in 140 g of water was added at 62 grams per hour. When all of the additions were complete, the latex was stirred under nitrogen for an additional 30 minutes and decanted to cool.

Table 1. Acrylic Latex Formulations

Ingredient	Baseline Latex	Experimental Latex
Methyl methacrylate	29.3	30.3
Butyl acrylate	23.9	24.9
Acrylic Acid	1.1	0
Aqueous AMPS Sodium Salt (active part)	0	0.6
Sodium Lauryl Sulfate	0.27	0.03
Synperonic® NP20	0.80	0
Sodium Bicarbonate	0.11	0.11
Sodium Persulfate	0.32	0.33
Total Water	44.2	43.8
Total Ingredients	100	100

Table 2. Vinyl Acrylic Latex Formulations

Ingredient	Baseline Latex	Experimental Latex
Vinyl Acetate	37.8	37.8
Butyl acrylate	9.46	9.46
Sodium Vinyl Sulfonate (active part)	0.47	0
Aqueous AMPS Sodium Salt (active part)	0	0.47
Sodium Lauryl Sulfate	0.95	0.95
Sodium Bicarbonate	0.14	0.14
Potassium Persulfate	0.21	0.21
Sodium Formaldehyde Bisulfite	0.02	0.02
t-Butylhydroperoxide	0.02	0.02
Total Water	50.87	50.87
Total Ingredients	100	100

Experimental Styrene Acrylic Latex Preparation. A solution of sodium lauryl sulfate (0.364 g) and sodium bicarbonate (1.4 g) in 395 g of water was prepared. Purge the solution subsurface with nitrogen to remove oxygen. A mixture of styrene (36.0 g, 0.346 moles) and butyl acrylate (32.6 g, .254 moles) and a 50% aqueous solution of AMPS sodium salt (1.4g, 3.06 mmol) were added as a separate feeds to the aqueous solution. When the temperature returned to 80°C, a solution of sodium persulfate (0.70 g, 2.94 mmol) in 10 grams of water was added. After 30 minutes, a mixture of styrene (324 g, 3.114 moles) and butyl acrylate (293.4 g, 2.286 moles) was added at a rate of 279 grams per hour. A 50% aqueous solution of AMPS sodium salt (12.6g, 27.5 mmol) in 140 grams of water was added at 68 grams per hour as a separate feed. After 30 minutes of monomer feed, a solution of sodium persulfate (3.5 g, 14.7 mmol) in 140 g of water was added at 62 grams per hour. When all of the additions were complete, the latex was stirred under nitrogen for an additional 30 minutes and decanted to cool. Table 3 compares the baseline and the experimental styrene acrylic formulations.

Paint preparation. Interior eggshell paints were prepared from the acrylic and the styrene acrylic latices in a manner typical of the decorative paint industry reflecting commonplace commercial practices and basic ingredients. The paints had a pigment volume concentration of 37% and a total solids content of 53%. A silk paint was prepared from a vinyl acrylic latex in a manner typical of the decorative paint industry reflecting commonplace commercial practices and basic ingredients. The paint had a pigment volume concentration of 35% and a total solids content of 45%.

Results and Discussion

The acrylic latex properties, Table 4, and the styrene acrylic latex properties, Table 5, showed a similar response to the use of AMPS sodium salt which is distinctly different than the response of the vinyl acrylic latex, Table 6. Acrylic and styrene acrylic latices containing AMPS sodium salt had significantly better stability towards cations and more than a 50% reduction in grit. The experimental acrylic latex, which was prepared with almost no surfactant, had a larger latex particle size than the baseline and correspondingly the Brookfield viscosity and the mechanical stability were reduced. If some sodium lauryl sulfate had been retained in the experimental formulation, it is expected that the particle size, Brookfield viscosity and mechanical stability would be closer to the baseline formulation values. The reduction in freeze/thaw stability is likely the absence of the non-ionic surfactant. The experimental styrene acrylic recipe retained the surfactant used in the baseline and both the latex particle size and Brookfield viscosity were unchanged. The mechanical stability was improved over the baseline. The experimental vinyl acrylic latex had slightly smaller latex particles, less grit and a higher Brookfield viscosity than the baseline; but the most significant improvement was in mechanical stability. All six of the latices were formulated into paints and the scrub resistance measured for each, Table 7. The baseline acrylic was good for only 775 cycles, but the experimental acrylic was good for 3500 cycles. The baseline styrene acrylic lasted longer, 3820 cycles, than the experimental acrylic latex but the experimental styrene acrylic was

Table 3. Styrene Acrylic Latex Formulations

Ingredient	Baseline Latex	Experimental Latex
Styrene	25.9	25.9
Butyl acrylate	23.4	23.4
Methacrylic Acid	0.5	0
Aqueous AMPS® Sodium Salt (active part)	0	0.5
Sodium Lauryl Sulfate	0.026	0.026
Sodium Bicarbonate	0.1	0.1
Sodium Persulfate	0.3	0.3
Total Water	49.7	49.8
Total Ingredients	100	100

Table 4. Acrylic Latex Properties

	Baseline Latex	Experimental Latex
Nonvolatiles (%),ASTM D 4758	49.30	50.14
Cation Stability: ml of 5% $CaCl_2$	1	>40
pH	4.5	6.9
Particle size (nm)	140	220
Polydispersity coefficient	0.107	0.069
Grit (ppm), ASTM D 5097	470	210
Minimum film forming °C, ASTM D 2354	17-18	17-18
Mechanical stability, (min to 5 poise)	>10 min	3.42 min
Freeze/thaw stability (cycles to fail), ASTM D 2243	>5	1
Brookfield cP, Spindle R2 @ 20 rpm	240	74

Table 5. Styrene Acrylic Latex Properties

	Baseline Latex	Experimental Latex
Nonvolatiles, % ASTM D4758	50.39	50.41
Cation Stability: ml of 5% $CaCl_2$	1	>25
pH	7.19	6.9
Particle size (nm)	281	280
Polydispersity coefficient	0.091	0.102
Grit (ppm) ASTM D5097	300	120
Minimum film forming °C ASTM D2354	23-24	20-21
Mechanical stability, (min to 5 poise)	4.70	8.35
Freeze/thaw stability (cycles to fail) ASTM D2243	1	1
Brookfield Vis., Spindle R2 @ 20rpm	55	59

Table 6. Vinyl Acrylic Latex Properties

	Baseline Latex	Experimental Latex
Nonvolatiles (%), ASTM D4758	49.10	48.58
Cation Stability: ml of 5% $CaCl_2$	18	17
pH	5	4.8
Particle size (nm)	147	132
Polydispersity coefficient	0.05	0.13
Grit (ppm), ASTM D5097	90	60
Minimum film forming °C, ASTM D2354	9-10	9-10
Mechanical stability, (min to 5 poise)	Coagulated @4minutes	>10 minutes
Freeze/thaw stability (cycles to fail), ASTM D2243	1	1
Brookfield cP, Spindle R2 @ 20rpm	54	170

Table 7. ASTM D 2486 Paint Scrub Resistance Comparisons

Paint	Scrub cycles
Acrylic baseline (eggshell)	775
Acrylic experimental (eggshell)	3500
Vinyl acrylic baseline (silk)	655
Vinyl acrylic experimental (silk)	687
Styrene acrylic baseline (eggshell)	3820
Styrene acrylic experimental (eggshell)	5700

even better at 5700 cycles. Both the vinyl acrylic baseline and the experimental paints gave poor performance.

The properties that AMPS appears to impart to a latex formulation suggest that AMPS performs two different roles. Although AMPS is certainly hydrophilic, it is not a surfactant. However, AMPS is capable of copolymerizing with hydrophobic monomers to form polymers which can function as surfactants. Gibbs et al [3] showed that a copolymer of 1 mole of AMPS and 6 moles of methyl methacrylate was an effective polymeric surfactant for the emulsion polymerization of styrene and styrene/butyl acrylate. Independently, Peiffer et al [4,5] showed that AMPS/styrene copolymers containing 40% to 60% AMPS behave as polymeric surfactants. Shild et al [6] studied the effect of AMPS on polystyrene coagulum, particle size, surface charge density, and latex stability in detail. It was also concluded that the formation of a polyelectrolyte from the homopolymerization of AMPS and its absorption on the surface of latex particles was possible and could contribute to latex stability. Corner[7] investigated the stabilization of polystyrene latices by polyelectrolytes, including AMPS homopolymer. The *in situ* formation of an AMPS homopolymer during emulsion polymerization seems unlikely unless the rate of AMPS addition is fast. The formation of copolymer polyelectrolytes seems much more likely. These copolymers, which will contain both nonionic hydrophobic and anionic hydrophilic segments may be more effective protective colloids than AMPS homopolymer. Recently, a vinyl acrylic latex process[8], a butadiene/styrene latex process[9] and an acrylic latex process[10] using AMPS sodium salt have been patented.

Conclusion

It has been demonstrated that AMPS® Monomer can be an effective partial replacement for surfactants and other water-soluble monomers in three common types of latex formulations: acrylic, vinyl acrylic and styrene acrylic. Improved latex properties and improved scrub resistance of coatings prepared from these latices have been demonstrated.

References
1. AMPS is a trademark of The Lubrizol Corporation.
2. Synperonic is a trademark of The ICI Corporation.
3. US 3,965,032 assigned to The Dow Company.
4. Peiffer, D. G.; Kim, M. W.; Kaladas, J. Polymer 1988, 29, 716-723.
5. Wang, Z.; Wang, J.; Chu, B.; Peiffer, D. G. J. Polym. Sci.: Part B: Polym. Phys. 1991, 29, 1361-1371.
6. Schild, R. L.; El-Aasser, M. S.; Poehlein, G. W.; Vanderhoff, J. W. Emulsions, Latices, Dispersions 1978, 99-128 Edited by Becher, P; Yudenfreund, M. N.; Marvin N. Dekker; New York, N.Y.
7. Corner, T.; Colloids and Surfaces 1981, 3, 119-129.
8. US 4,812,510 assigned to The Glidden Company.
9. US 5,274,027; US 5,302,655 assigned to The Dow Chemical Company.
10. EP 770,655 A2, assigned to Rohm and Haas Company.

Chapter 6

Photoinitiated Polymerization of a Semifluorinated Liquid Crystalline Monomer Exhibiting a Reduced Rate in the Smectic Phase

C. E. Hoyle, L. J. Mathias, C. Jariwala, D. Sheng, and P. E. Sundell

Department of Polymer Science, The University of Southern Mississippi, Box 10076, Hattiesburg, MS 39406–0076

The photoinitiated polymerization of a sterically hindered semi-fluorinated monomer, which is characterized by a hindered radical chain propagation site, was followed both in an isotropic and a highly ordered smectic liquid crystalline phase. The polymerization rate is slower in the smectic phase than in the isotropic phase, presumably a result of a decrease in both the propagation and termination rate constants. The decrease in the propagation rate results in a very slow persistent increase in polymer molecular weight after the initiating light source is removed. Polymerization from the smectic phase of the monomer proceeds in a non-equilibrium matrix.

Over the past two decades, the free radical polymerization of liquid crystalline monomers has been actively investigated. Questions dealing with the effect of nematic and smectic ordering on the polymerization rate, the molecular weight of the polymers produced (for monofunctional monomers), and the retention of order in the crosslinking polymerization of multifunctional monomers have been addressed (1-19) in a number of publications. Based on a composite of results from our laboratory (13, 14) as well as reports in references 8 and 9, we conclude that in general nematic or cholesteric nematic structure, which involves only directional ordering of the monomer groups participating in the polymerization process, has relatively little effect upon free-radical polymerization kinetics, the chemical structure, or molecular weight distribution of the polymers produced. Polymerization in smectic phases, however, can have a significant effect. Earlier studies in the literature dealing with the kinetics of polymerization in smectic phases are summarized in four review articles (8,18-19). In recent results from our laboratory (16-17), we have found that polymerization of the cholesterol bearing methacrylate monomer CMA-10 in the smectic A phase, and the semifluorinated methacrylate F12H10MA and acrylate F12H10A monomers in the smectic phase (smectic B or closely related type), exhibit significant rate enhancements compared to the isotropic phases. The rate increases are especially high in the smectic phases of F12H10MA and F10H12A. In the case of the CMA-10 monomer, we have shown (17) via simultaneous light scattering and exotherm rate measurements (both steady-state and transient conditions) that polymerization proceeds in a single smectic phase in which termination is lowered by about three orders of magnitude compared to polymerization in the isotropic phase. The propagation rate constant for CMA-10 polymerization, by contrast, is relatively unaffected by the phase of the medium since propagation is apparently controlled by a chemical rather than a diffusion controlled rate limiting process. Although we have not yet measured the individual rate constants for propagation and termination of F12H10MA and F12H10A in the smectic phase, we have tentatively projected (based on our experience with CMA-10) that a decrease in the termination rate constant is responsible for the unprecedented overall increases in polymerization rate.

In this chapter, we present our results for the free-radical polymerization of a semifluorinated monomer designated F10. A comparative evaluation of polymerization rates in the smectic and isotropic phases of F10 is given. The resulting effect on the relative polymer molecular weight distributions is critically analyzed with respect to the long lived free-radical polymer chain ends in the smectic phase. Since this monomer is sterically crowded at the terminal radical site and exhibits a highly viscous smectic phase at room temperature, we might expect to see a rate phenomenon which reflects both lowered termination and propagation rate constants for polymerization in the liquid crystalline phase.

Experimental

The synthesis of the F10 monomer has been described in a previous publication (20). The exotherms generated from the photoinitiated polymerization of F10 monomers with one weight percent photoinitiator were recorded on a Perkin-Elmer 2B calorimeter in an isothermal mode (non-scanning). The DSC was modified to allow penetration of the light from a medium pressure mercury lamp source. The molar heat of polymerization for the F10 monomer, 13.76 kcal/mol, was obtained by correlating exotherm data with direct monomer conversion by HPLC analysis. Simultaneous recording of the exotherm curve and laser light scattering was obtained by reflecting a HeNe laser source off the aluminum sample pan containing the sample. The

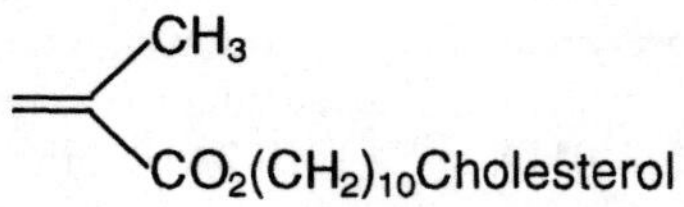

CH_3

$CO_2(CH_2)_{10}Cholesterol$

CMA-10

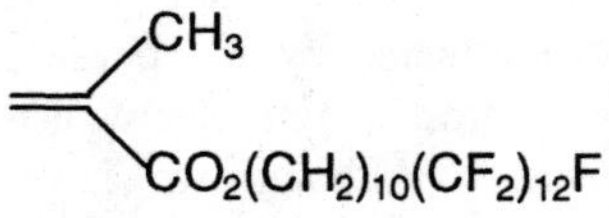

CH_3

$CO_2(CH_2)_{10}(CF_2)_{12}F$

F12H10MA

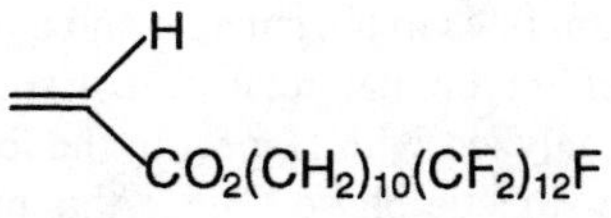

H

$CO_2(CH_2)_{10}(CF_2)_{12}F$

F12H10A

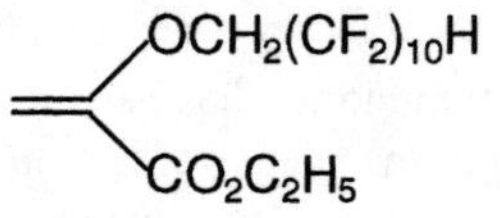

$OCH_2(CF_2)_{10}H$

$CO_2C_2H_5$

F10

mercury lamp source used to initiate polymerization could be attenuated via neutral density filters. The phase diagram was determined by both DSC scans and observation of thermally scanned samples with an optical microscope equipped with cross-polarizers. The texture of the monomer was identified by comparison with optical micrographs and accompanying descriptions in reference 21. ESR spectra were obtained at Louisiana State University. GPC chromatograms were obtained on a Waters GPC equipped with a refractive index detector. In cases where molecular weights are quoted, they are given with respect to polystyrene standards run on the same set of columns under the identical conditions used to evaluate experimental samples.

Results and Discussion

Although we have investigated and characterized the thermal behavior and polymerization kinetics of several α-methylene substituted semifluorinated methacrylates, we have chosen to highlight in this chapter the monomer designated as F10. The F10 monomer is particularly interesting since it exhibits a highly ordered smectic liquid crystalline phase at room temperature. Hence F10 is a convenient monomer to work with since high temperatures are not required in order to observe polymerization kinetics in the liquid crystalline phase. Following a brief thermal/microscopic description of F10, the photopolymerization kinetics are evaluated via exotherm analysis at temperatures corresponding to smectic and isotropic phases of the monomer.

A.   Thermal/Microscopic/X-ray Characterization of F10.

The smectic phase is present upon the initial casting of the sample at 20 °C into an indented DSC sample pan. As confirmed by cross-polarized optical microscopic investigation of F10 with one percent 1,1-dimethoxy-1-phenyl acetophenone (DMPA) photoinitiator, the first heating cycle is characterized by clearing from a smectic phase to an isotropic phase at 32 °C. Upon subsequent cooling at a rate of 10 °C/min, the smectic phase is obtained at temperatures below approximately 27 °C (actually the phase is once again fully developed below about 24 °C). The X-ray diffraction pattern of F10 at 20 °C clearly shows a well-defined inter-layer spacing of about 20 A° and an intermolecular spacing of 4.95 A°. The inter-layer spacing (20 A°) corresponds to the full length of the F10 molecule obtained by a molecular modeling calculation and is suggestive of a smectic B phase created by aggregation of the low surface tension fluorinated -$(CF_2)_{10}$H chains. Strong evidence for assignment of the liquid crystalline structure to a smectic B phase is obtained from the cross-polarized optical microscopic examination of F10 at temperatures below the phase transition (see Figure 1): The optical micrograph in Figure 1 obtained under cross-polarizers has the classic mosaic texture of a smectic B phase (22). Finally, we note that we did not observe

Figure 1.   Mosaic texture of F10 monomer observed at 20 °C under a
polarizing microscope.

crystallization of F10 at temperatures as low as 0 °C at the DSC scanning rate employed. Presumably, if the monomer is annealed at low enough temperatures, a slow crystallization might occur.

B.   Polymerization of F10--Continuous Light Source.

We now ask the question, what effect does the organization of the F10 monomer in the smectic phase have upon the polymerization kinetics initiated by light? For other liquid crystalline systems described in the introduction, we have obtained a variety of results with respect to the effect of liquid crystalline media on polymerization kinetics. In the present investigation, we examine the polymerization kinetics of F10 which is substituted with two quite large groups at the reactive chain end. The innate effect of such groups at the radical chain centered carbon is to lower the propagation rate process. With this in mind, the photopolymerization exotherm curves for F-10 will be considered, first as a means to follow the relative rates in the smectic phase, and second to establish the basic kinetic mechanism governing the free radical polymerization process.

Figure 2 shows the basic exotherms and corresponding percent conversion versus time plots for polymerization of F-10 with one percent photoinitiator (2,2-dimethoxyacetophenone) by a medium pressure mercury lamp at 20 °C (from the monomer smectic phase) and 35 °C (from the isotropic monomer phase). At temperatures greater than 35 °C, the exotherm peak maximum (not shown) for F10 decreases as the temperature at which the photopolymerization is conducted is increased. At temperatures less than 32 °C, the exotherms are markedly decreased and their shapes are significantly different from those at a temperature where the monomer is in an isotropic phase prior to the onset of polymerization. The results in Figure 2 are representative of exotherms recorded at temperatures above and below the phase transition of the monomer. A traditional Arrhenius analysis of the exotherm data for F10 is not possible since as noted there is a decrease in exotherms as the temperature increases in the isotropic phase (due to what is most likely a ceiling temperature effect). However, we point out that all exotherms recorded for samples at temperatures in the smectic phase (up to the clearing temperature) are much lower than the exotherm in the isotropic phase at 35 °C. We have chosen to present results at 20 °C as an example of polymerization in the smectic phase. Concerning the exotherm curve at 35 °C, and restricting the discussion to lower conversions (up to 35 percent), we note that the exotherm is markedly greater (at a given exposure time) for polymerization at 35 °C than at 20 °C. A semi-log plot (not shown herein) of the reciprocal of the monomer concentration versus the exposure time at low conversions for the data in Figure 2 (at 20 °C and 35 °C) is linear indicating that termination is a bimolecular process at both temperatures.

Returning to consideration of the full exotherm curves in Figure 2a, we note a sudden increase at about 35 percent conversion in the polymerization

a.

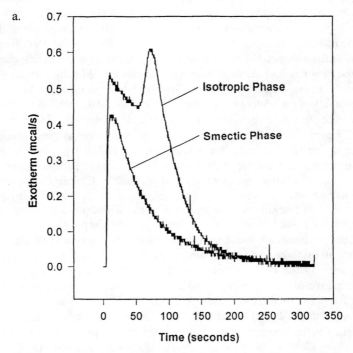

Figure 2.   (a) Polymerization exotherms of F10 at 20 °C and 35 °C initiated with medium pressure mercury lamp (0.44 mJ cm^{-2} sec^{-1}), and corresponding (b) percent conversion versus time plots for polymerization of F10 at 20 °C and 35 °C.

b.

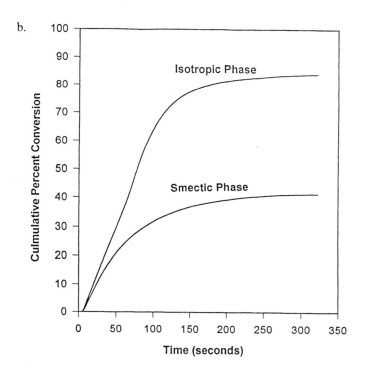

Figure 2. *Continued.*

rate at 35 °C. Subsequent to the rate acceleration, changes in the medium were determined both by direct observation (turbidity) as well as via cross-polarized optical microscopy. Interestingly, at 20 °C no abrupt rate increase occurs and no changes are detected in medium turbidity or optical texture by cross-polarized optical microscopy during the polymerization process. In the case of polymerization at 20 °C, any medium changes which occur are thus rather minor with respect to disruption of medium texture or sudden rate acceleration. A closer examination is warranted to substantiate the general observations with respect to rate acceleration and medium textural changes observed at 35 °C via optical microscopy.

In order to confirm the onset of the abrupt rate acceleration at 35 °C and to provide a more direct correlation with the medium phase changes, the DSC exotherm system was modified to allow a non-obtrusive cw HeNe laser (632 nm) to reflect from the sample pan coincident with the exposure of the sample to the medium pressure mercury lamp in the UV. Figures 3 and 4 show simultaneously the exotherm curve, cumulative percent conversion versus time plot, and corresponding reflected light intensities at 20 °C and 35 °C. For polymerization at 35 °C, it is apparent (Figure 4) that a decrease in reflected light due to scattering by the sample occurs after the rate acceleration begins. Moreover, for polymerization at 20 °C (Figure 3) in the smectic phase, no change in the light intensity reflected from the pan through the sample could be detected at any point during the polymerization. In other words, any change in the light scattered as a result of a change in the medium (phase separation, medium phase change, or both) is minimal. This corresponds well with direct observation of the polymerization medium as noted previously.

C.   Phase Diagram Comparison with Polymerization.

Polymerization in the smectic phase at 20 °C, as demonstrated by the laser light reflection investigation in Figure 3, results in no apparent major change in the medium phase character as polymerization progresses. One question which is particularly important deals with the nature of the medium produced upon polymerization of the smectic monomer at low temperature (20 °C) compared to a medium produced by cooling of an F10 monomer/polymer physical mixture from the melt. First, consider the phase diagram obtained via DSC/microstopic examination upon heating (second) of physical mixtures of F10 monomer/polymer (Figure 5). As already noted, the pure F10 monomer exhibited a smectic B phase below 32 °C. The crystalline melting point for the pure polymer was 60 °C. Except for compositions of greater than about 80-85 percent polymer, physical monomer/polymer sample mixtures exhibited two organized phases (crystalline, smectic) at lower temperatures. Upon heating samples with polymer content less than about 80-85 percent a smectic-to-isotropic phase transition occurred at temperatures ranging from 32 °C (pure monomer) to about 29 °C (74 percent polymer) followed by a crystalline-to-

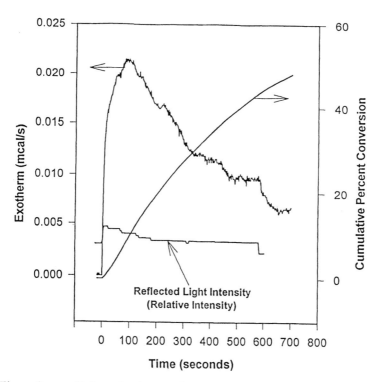

Figure 3.   Polymerization exotherm, percent conversion versus time plot, and corresponding HeNe light scattering for polymerization of F10 with a medium pressure mercury lamp (Intensity = 0.0079 mJ cm^{-2} sec^{-1}) at 20 °C.

64

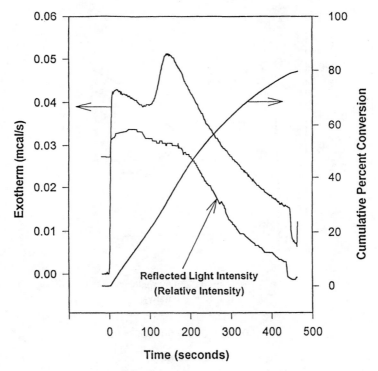

Figure 4.   Polymerization exotherm, percent conversion versus time plot, and corresponding HeNe light scattering for polymerization of F10 with a medium pressure mercury lamp (Intensity = 0.0079 mJ cm^{-2} sec^{-1}) at 35 ºC.

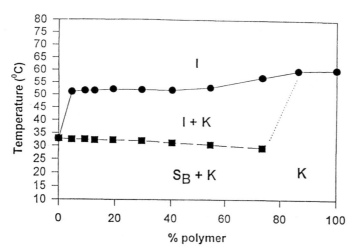

Figure 5.   Phase diagram of F10 monomer/polymer physical mixtures
            obtained upon heating of samples previously cooled from
            the melt to below room temperature.

isotropic transition at higher temperatures (ranging from about 52 °C for 5 percent polymer content to about 57 °C for 74 percent polymer content). These melting temperatures for the crystalline phase (detected by optical microscopy) in the mixed monomer/polymer systems were only modestly less than the crystalline melting point of the pure polymer (60 °C). (Incidentally, we could not detect any liquid crystalline phase for the polymer). Apparently, the crystalline region of the polymer does not readily incorporate monomer since the polymer melting temperature is only marginally decreased by the presence of the F10 monomer. In summary, the monomer/polymer mixtures (up to 80-85 percent polymer samples) have little tendency to form a single phase at low temperatures, but rather form two phases, presumably one dominated by the monomer and the other by the polymer.

When polymerization occurs from the smectic phase at 20 °C, the resultant media are quite different from those produced by simply cooling physical mixtures of the same monomer/polymer content from the melt. As an example, consider the medium produced by polymerization of F10 at 20 °C to 31.6 percent conversion. If this sample is heated in the DSC, two very broad (about 10 °C) endothermic transitions are recorded with peak maxima at about 42 °C and 53-55 °C (broad--more than one endotherm peak). If the sample is cooled from the melt and once again reheated, the endotherms have peak maxima at approximately 32 °C and 53 °C (narrow--appears to be one endotherm peak) and the scan is essentially identical to that expected from a physical mixture. We thus conclude that the medium generated upon polymerization at 20 °C is not equivalent thermodynamically to the medium obtained upon cooling of the sample for the melt or for a physical mixture representing the thermodynamic equilibrium state. The temperature (peak maximum at 42 °C) recorded for the first transition may represent melting of a phase(s) comprised of high monomer content with enhanced order resulting from a non-thermodynamic-equilibrium mixing with the polymer that is formed initially upon polymerization.

D.   Shuttered Light Exposure Sequence (ESR and Exotherm Results).

Up until this point, we have determined that the photoinitiated polymerization of the F10 monomer in the smectic phase is slower than in the isotropic phase when the continuous output of a mercury lamp is used as the initiating light source. The consequences and origin of this rate phenomenon will be further explored with respect to dark polymerization in the smectic phase after the initiating light source is suddenly terminated. The results will provide direct evidence via exotherm and ESR analysis of the long-lived polymer radical chains in the smectic phase, as well as an estimate of the termination/propagation rate constant ratio in the smectic phase.

Figure 6 shows exotherms for polymerization of F10 at 35 °C and 20 °C

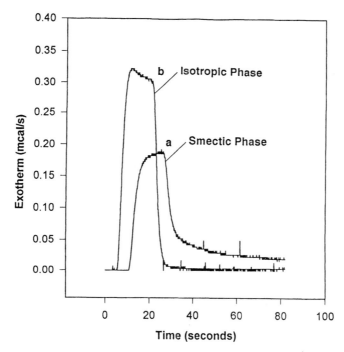

Figure 6.   Exotherm decays for F10 polymerization initiated with mercury lamp source for (a) 7.2 secs at 20 °C (5 percent conversion attained before shuttering lamp), and (b) 15.3 secs at 35 °C (9 percent conversion attained before shuttering lamp).

initiated by exposure of the sample for various time periods followed by an abrupt termination of the source via a rapid shutter placed between the lamp source and the DSC chamber cell. At 35 °C, the exotherm rapidly returns to the baseline after the light source is terminated. By contrast, at 20 °C in the smectic B phase the polymerization continues for a long period of time after the light source is removed. By making a plot of $[M]/[-d[M]/dt]$ versus time for the data in Figure 6a (assuming a boundary condition of time = 0 at the instant that the shutter closes), the slope yields a value of ~50 for k_t/k_p according to equation 1(23),

$$[M]/[-d[M]/dt] = (k_t/k_p)t + [M]_o/[-d[M]/dt]_o \qquad \text{eq. 1}$$

where $[M]_o$ is the monomer concentration at time zero, $[M]$ is the monomer concentration at time t after the lamp is terminated, $[-d[M]/dt]_o$ is the polymerization rate at time zero, and $[-d[M/dt]$ is the polymerization rate at time t after the lamp is terminated. The k_t/k_p value is quite low compared to conventional values typically obtained for free radical polymerization of isotropic monomers (see Table I for listing of values for several monomers). As well, the k_t/k_p ratio at 35 °C in the isotropic phase at low conversion is estimated to have a lower limit value of 1000 from the exotherm decay in Figure 6b. No doubt a drastic decrease in the value of k_t, relative to k_p, must be a direct consequence of the reduction in medium mobility in the smectic phase compared to the isotropic phase.

We have indeed shown (17) that k_t is markedly reduced (up to 3 orders of magnitude) for the CMA-10 monomer resulting in a substantial rate enhancement since k_p was independent of the phase. Although for F10 we are not able to calculate actual rate constants in the isotropic and smectic phases, we feel confident in assuming that k_t also decreases substantially in the smectic phase compared to the isotropic phase. We are then faced with a dilemma; if k_t for F10 is very low in the smectic phase at 20 °C, then why is the steady-state rate lower for F10 in the smectic phase than in the isotropic phase? Why don't we observe the same type of rate enhancement (or even greater) in the present case for F10 in the smectic phase? In the previous case for the smectic phase of CMA-10, k_p was, except for a temperature effect predictable by a simple Arrhenius relationship, unaffected by the medium order; i.e., the magnitude of k_p was not affected by the smectic liquid crystalline phase. In the present case for F10 in the smectic medium, we suspect that k_p, as well as k_t, decreases in comparison to the isotropic phase. This, of course, would suggest that in the smectic phase k_p is governed by diffusion, not chemical, control. Such a result might be expected for a monomer such as F10 where the reactive radical chain is sterically hindered and the medium provides for limited mobility. The extremely viscous nature of F10 at 20 °C in the smectic phase indeed suggests a medium capable of rendering diffusion controlled propagation, especially if the chain end is sterically crowded.

What other observations might be made with respect to the very long-lived nature of the F10 polymer radical chains? First, we present in Figure 7 a series of ESR spectra obtained at four times corresponding to a 1200 second exposure of an F10/photoinitiator mixture at 20 °C to a medium pressure mercury lamp: (a) Immediately after the 1200 second exposure, (b) 682 seconds after the initial 1200 second exposure, (c) 7500 seconds after the 1200 second exposure, and (d) 7500 seconds after the 1200 second exposure followed by raising of the sample temperature to 80 °C. It is obvious that the ESR signal generated by photolysis for 1200 seconds is quite large (Figure 7a) and persists well after (Figure 7b and 7c) the light exposure is terminated, provided the sample temperature is maintained at 20 °C. The shape of the ESR signal is that expected for the growing radical chain end of the F10 polymer, and is consistent with the long lived polymerization process as identified by the exotherm in Figure 6a obtained after the light source was removed. It appears as though a long lived free radical type chain process follows when polymerization is initiated in the smectic phase of the F10 monomer. Only after the temperature of the medium is raised to 80 °C (Figure 7d), well above the mixture isotropization temperature, is the radical signal lost, indicating a rapid termination process afforded by a more "mobile" medium (high diffusivity) at the higher temperatures. Incidentally, when polymerization was initiated at 35 °C in the isotropic phase, we observed no ESR signal at least up to the instant that the medium acceleration depicted in Figure 2a occurred. This presumably results from a very low steady-state radical concentration in the isotropic phase. We conclude from the ESR results for F10 photopolymerization at 20 °C that radical chains continue to exist at long times after the initiating light source is removed from the sample.

A natural consequence of the long lived radical chain is continuation of the propagation process after the light source is terminated: This was clearly indicated by the exotherm in Figure 6a. In an additional experiment, a nitrogen purged sample of F10 with one percent DMPA photoinitiator was exposed to a medium pressure mercury lamp (pyrex filter) to give a low percent total conversion after the exposure was terminated via a shutter. The sample was then immediately exposed to oxygen to terminate chain propagation and subsequently injected into a GPC for analysis. The identical procedure was repeated for two additional samples except that after the light source was terminated delay time periods of two minutes and ten minutes were imposed before exposing the samples to oxygen. The intent of this set of experiments was to determine if the molecular weight of the polymer samples produced by polymerization at 20 °C continues to increase even after the initiating light source is removed. Figure 8 illustrates that indeed the molecular weight distribution of the polymer sample continues to increase well after the light source is terminated; i.e., polymer chains growing at the time that the light source is removed continue to increase in molecular weight. These results are certainly in accord with the exotherm decays in Figure 6a and the ESR spectra

70

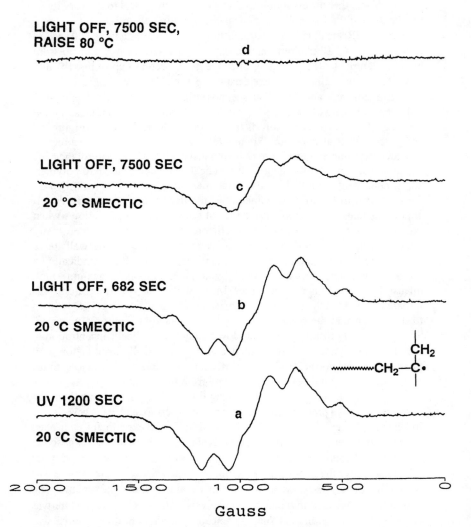

Figure 7.   ESR spectra of F10 polymer radical chain for polymerization at 20 °C upon a 1200 second exposure to a medium pressure mercury lamp: (a) Immediately after the 1200 second exposure at 20 °C, (b) 682 seconds after the initial 1200 second exposure at 20 °C, (c) 7500 seconds after the initial 1200 second exposure at 20 °C, and (d) 7500 seconds after the initial 1200 second exposure at 20 °C followed by raising of the sample temperature to 80 °C.

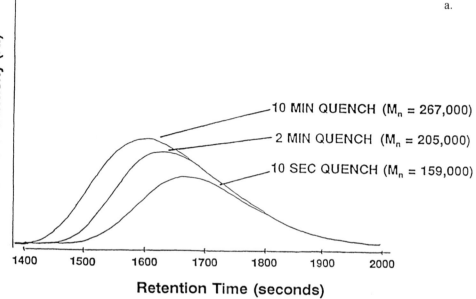

a.

Figure 8.   (a)   GPC chromatograms and corresponding (b) Cumulative weight fraction versus log molecular weight for polymer attained by photoinitiated polymerization for various dark periods (after light removal and identified on figure) followed by exposure of sample to oxygen (quenching). Delay times after termination of the light source and before oxygen quenching are shown on figure, as well as molecular weights (compared to polystyrene standards) at the peak maximum. In each case the percent monomer conversion is less than 10 percent.

Continued on next page.

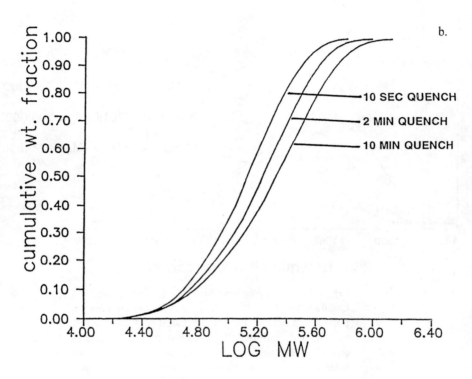

Figure 8. *Continued.*

in Figure 7 and lend credence to the very long lived nature of the polymer radical chains for polymerization in the smectic phase at 20 °C. Incidentally, there was no change in the molecular weight of polymer samples generated at 35 °C, even if long times were imposed after the initiating light source was removed before the sample was exposed to oxygen. This occurs since F10 polymerization at 35 °C exhibits relatively rapid chain termination and cessation of the polymerization after removal of the initiating light source.

Conclusions

In this paper, we have described the characterization of a polymerization process which proceeds at lower rates in the liquid crystalline smectic phase than in the isotropic phase. The polymerization in the highly organized and viscous smectic phase, for which the propagating radical is highly hindered, proceeds via a steady-state process in which the k_t/k_p ratio is quite low. Our results strongly suggest that both k_t and k_p are severely reduced in the ordered smectic phase, and that k_p, as well as k_t, is diffusion controlled. Exotherm decay curves, ESR spectra of the polymer radical chains, and the increase in polymer molecular weight in the absence of the initiating light source attest to the very long-lived nature of the polymer radical chains. The results of rate dependence on phase structure presented in this paper for polymerization of F10 in the smectic phase at temperatures near room temperature are quite different from the results for polymerization of the semifluorinated acrylate F12H10A and methacrylate F12H20MA in the smectic B phase presented previously (15, 16). This is at least partially due to the sterically crowded nature of the propagating radical.

Acknowledgments

We acknowledge multiple grants from the National Science Foundation [DMR] for support of this work. We also thank Dan Church at Louisiana State University for assistance in ESR measurements.

Literature Cited

1. Broer, D. J.; Mol, G. N. ; Challa, G. *Makromol. Chem.* **1989**, 190, 19.
2. Broer, D. J.; Mol, G. N. *Makromol. Chem.*, **1991**, 192, 59.
3. Broer, D. J.; Hikmet, R. A. M.; Challa, G. *Makromol. Chem.*, **1989**, 190, 3201.
4. Broer, D. J.; Boven, J.; Mol, G. N.; Challa, G. *Makromol. Chem.*, **1989**, 190, 2255.
5. Broer, D. J.; Lub, J.; Mol, G. N. *Macromolecules*, **1993**, 26, 1244.
6. Hikmet, R. A. M.; Lub, J.; Brink, P. Massen vd. *Macromolecules*, **1992**, 25, 4194.

7.   Hsu, E. C. ; Blumstein, A. *J. Polym. Sci., Polym. Lett. Ed.*, **1977**, 15, 129.

8.   Blumstein, A. Mid. *Macromol. Monogr.* **1977**, 3, 133.

9.   Paleos, C. M. Labes, M. M. *Mol. Cryst. Liq. Cryst.* **1970**, 11, 385.

10.   Hikmet, R. A. M. *Macromolecules*, **1992**, 25, 5759.

11.   Broer, D. J.; Heynderickx, I. *Macromolecules*, **1990**, 23, 2474.

12.   Hikmet, R. A. M.; Broer, D. J. *Polymer*, **1991**, 32, 1627.

13.   Guymon, C. A.; Bowman, C. N. *Macromolecules*, **1997**, 30, 5271.

14.   Hoyle, C. E.; Chawla, C. P.; Kang, D.; Griffin, A. C. *Macromolecules*, **1993**, 26, 758.

15.   Hoyle, C. E.; Kang, D. *Macromolecules*, **1993**, 26, 844.

16.   Hoyle, C. E.; Kang, D.; Williamson, S. *Macromolecules*, **1996**, 29, 8656.

17.   Hoyle, C. E.; Watanabe, T. *Macromolecules*, **1994**, 27, 3790.

18.   (a) Percec, V.; Johnson, H.; Tamasos, D. in *Polymerizations in Organized Media*, Paleos, C. M., Ed.; Gordon and Breach Publishers: Philadelphia, **1992**; p 1.; (b) Paleos, C. M. *Chem. Soc. Rev.*, **1985**, 14, 45.

19.   Barrell, E. M.; Johnson, J. F. *J. Macromol. Sci., Rev. Macromol. Chem.*, **1979**, 17, 137.

20.   Jariwala, C. P.; Sundell, P.-E.; Hoyle, C. E.; Mathias, L. J.; *Macromolecules*, **1991**, 24, 6352.

21.   Demus, D.; Richter, L. *Textures of Liquid Crystals*, Verlag Chemie, Weinheim, N. Y. **1978**.

22.   Odian, G. *Principles of Polymerization*, John Wiley and Sons, New York, **1991**, 275.

23.   Tryson, G.R.; Shultz, W.R. *J. Polym. Sci. Polym. Phys. Ed.*, **1979**, 17, 2059.

MATERIALS

Chapter 7

Single-Molecular Assemblies of Hydrophobically-Modified Polyelectrolytes and Their Functionalization

Yotaro Morishima and Akihito Hashidzume

Department of Macromolecular Science, Graduate School of Science, Osaka University, Toyonaka, Osaka 560–0043, Japan

Hydrophobic associations in random copolymers of sodium 2-(acrylamido)-2-methylpropanesulfonate and some methacrylamides and methacrylates substituted with bulky hydrophobes are described with a focus on preferential intrapolymer self-association which leads to the formation of single-macromolecular assemblies (i.e., unimolecular micelles). Structural parameters that critically determine the type of the macromolecular association (i.e., intra- vs. interpolymer associations) are discussed, which include the type of hydrophobes, their content in a polymer, sequence distribution of electrolyte and hydrophobic monomer units, and the type of spacer bonding. Functionalization of single-macromolecular assemblies with some photoactive chromophores is also presented.

Macromolecular self-assembling phenomena are of great current scientific and technological interest, because these phenomena are relevant to molecular organization in biological systems and also to various practical applications (1–3). Macromolecular self-assemblies can be driven by non-covalent interactions including Coulombic, hydrogen bonding, van der Waals, and hydrophobic interactions. Among others, hydrophobic interactions are a major driving force for the self-organization of amphiphilic polymers in water.

A primary approach to the architecture of self-assembling macromolecules is to covalently introduce hydrophobes into water-soluble polymers. A predetermined number of hydrophobes can be introduced into a polymer chain by copolymerization of hydrophilic and hydrophobic monomers with a block, alternating, or random sequence distribution.

The association of polymer-bound hydrophobes can occur either within a single polymer chain or between different polymer chains, or both at a time, depending on the type of amphiphilic polymer. In highly dilute aqueous solutions, in general, hydrophobic associations may occur within a polymer chain, but with an increase in the polymer concentration, a tendency for interpolymer association may increase. In the case of hydrophobically-modified nonionic polymers, the association of polymer-bound hydrophobes can occur even if the hydrophobe content in a polymer is very low (4–6). In the case of hydrophobically-modified polyelectrolytes, however, a relatively high content of hydrophobes is necessary for the association to occur, because hydrophobic interactions compete with electrostatic repulsions. Thus, the balance of the contents of hydrophobic and charged units in a polymer is a critical factor for the hydrophobic association to occur. For example, amphiphilic polycarboxylic acids, which adopt an extended chain conformation at high pH, would collapse into a compact conformation upon decreasing pH (7–10). This transition from an extended to a compact structure, a typical of cooperative processes that can be viewed as a two-state transition, is abrupt enough to define a critical transition pH.

A large number of experimental results reported so far on hydrophobically-modified polyelectrolytes indicate that the type of hydrophobes, their content in a polymer, the sequence distribution of electrolyte and hydrophobic monomer units in a polymer chain, and the type of spacer bonding between hydrophobes and the polymer chain are some of the structural parameters that determine whether intra- or interpolymer hydrophobic association occurs preferentially (11–20). Some of the structural parameters are primarily important for the hydrophilic-hydrophobic balance in a polymer. For example, the maximum amount of a hydrophobe that can be incorporated into a polymer, while keeping the polymer soluble in water, depends on the type of the hydrophobe, its content, and the spacer bonding. A subtle difference in the sequence distribution in random copolymers has a large effect on the hydrophobic association (11); hydrophobic block sequences have a strong tendency for interpolymer association, whereas random and alternating sequences have a tendency for intrapolymer association. Because the polymer chain exerts steric constraints to polymer-bound hydrophobes, the degree of the motional and geometrical freedom of polymer-bound hydrophobes has an important effect on their self-association. Therefore, the spacing between hydrophobes and the polymer backbone is a key element to control the hydrophobic association (20).

Random copolymers of sodium 2-(acrylamido)-2-methylpropanesulfonate (AMPS) and N-dodecyl-, N-cyclododecyl-, or N-adamantylmethacrylamide are soluble in water up to about 60 mol % of the hydrophobic methacrylamide content (19,21–22). In contrast, random copolymers of AMPS and dodecylmethacrylate are soluble in water only when the content of dodecylmethacrylate is ≤10 mol % (23). These results indicate that there is a great difference in the solubility in water between the polymers with amide and ester spacer bonds connecting hydrophobes to the main chain. Both the amide-spacer and ester-spacer polymers show a tendency for interpolymer association when the hydrophobe contents are lower than about 10 mol %. This tendency is much more pronounced in the ester-spacer polymers than in the amide-spacer polymers. As the hydrophobe contents in the polymers are increased up to about 20 mol % or higher, the amide-spacer polymers show a strong preference for intrapolymer self-association even in a concentrated regime (24). On the other hand, the ester-spacer polymers give strongly turbid solutions when the hydrophobe content is increased to about 15 mol % (23).

This chapter will discuss hydrophobic associations in random copolymers of AMPS and some hydrophobic methacrylamide and methacrylate comonomers with a focus on the intra- versus interpolymer self-association in connection with the type of hydrophobes, their content in the polymers, and spacer bonding. A particular emphasis will be placed on intrapolymer association of hydrophobes which leads to single-molecular self-assemblies. Functionalization of the single-macromolecular assemblies with some photoactive chromophores will also be presented briefly.

General Considerations of Hydrophobic Associations in Amphiphilic Random Copolymers.

In aqueous solutions of hydrophobically-modified water-soluble polymers, strong interpolymer hydrophobic association often leads to gelation or precipitation. It is well-known, however, that **AB**- and **ABA**-type block copolymers, where **A** and **B** represent hydrophilic and hydrophobic blocks, respectively, form micelles with a hydrophobic core and a hydrophilic corona (25). The formation of these polymer micelles is described by a closed association process (26–28), and at a thermodynamic equilibrium, the aggregation number is determined by the minimum free energy. Some of the examples of such block copolymers include poly(ethylene oxide)-*block*-poly(propylene oxide) (*29–33*), polystyrene-*block*-poly(methacrylic acid) (*34–36*), and polystyrene-*block*-poly(acrylic acid) (*37*).

In contrast to the di- and triblock copolymers, multiblock copolymers, in which two or more hydrophobic block sequences are linked with hydrophilic block sequences in a polymer chain, may undergo multimolecular aggregation where the aggregation number is not limited. This type of interpolymer hydrophobic association, which may be termed "open association" in contrast to the closed association, normally leads to bulk-phase separation or gelation. As the number of hydrophobic sequences in a multiblock copolymer increases, intra- and interpolymer hydrophobic associations are likely to occur concomitantly. This is particularly true for amphiphilic random copolymers where hydrophobic and hydrophilic units are randomly distributed on a polymer chain (Figure 1). If, however, intrapolymer hydrophobic association occurs preferentially, a single-molecular self-assembly may lead to the formation of a unimolecular micelle. When the content of hydrophobes in a random copolymer is sufficiently low, a "flower-like" unimolecular micelle may be formed, which consists of a hydrophobic core surrounded by hydrophilic loops shaped into "petals" (*38,39*). This flower-like micelle can be viewed as a secondary structure of a polymer chain. As the content of hydrophobes in a copolymer is increased, the size of the hydrophobic core increases and in turn the size of the hydrophilic petals decreases. Accordingly, the secondary structure may become unstable because a significant portion of the surface of the hydrophobic core is exposed to water, thus leading to a further collapse of the secondary structure into a tertiary structure due to the association of hydrophobic cores (Figure 1). The tertiary structure is made up with a number of flower-like micelles collapsed into a compact assembly.

Hydrophobic Association of Random Copolymers of AMPS and Cholesterol-Bearing Methacrylates.

Cholesterol plays an important role in controlling membrane fluidity in biological systems, arising from its strong tendency for hydrophobic interactions and the rigidity of the steroid ring (*40*). Amphiphilic derivatives of cholesterol are known to form various types of molecular assemblies in aqueous solution such as lyotropic liquid-crystalline phases, liposomes, micelles, and ordered monolayers (*41*). A polysaccharide (pullulan) covalently incorporated with a few cholesterol moieties per 100 glucose units forms stable nanoparticles with the diameters of 20–30 nm in water via interpolymer cholesterol associations, indicating that cholesterol has a strong tendency for self-association even if the cholesterol contents in polymers are very low (*42,43*).

The two types of cholesterol-carrying polyelectrolytes (Chart 1), copolymers of AMPS with cholesteryl methacrylate (CholMA) and with cholesteryl 6-methacryloyloxyhexanoate (Chol-C5-MA), were synthesized (*20*). In the former, cholesterol is directly attached to the main chain via an ester bond, whereas in the latter, cholesterol is linked to the main chain via a pentamethylene spacer connected with two ester bonds. The associative behavior of these cholesterol-carrying polymers in aqueous solution was investigated by turbidimetry, ^{1}H NMR, nonradiative energy transfer (NRET), size exclusion chromatography (SEC), quasielastic light scattering (QELS), static light scattering (SLS), and viscometry (*20*). For NRET studies, fluorescence-labeled terpolymers of AMPS, CholMA or Chol-C5-MA, and *N*-(1-pyrenylmethyl)methacrylamide or *N*-(1-naphthylmethyl)methacrylamide were employed (Chart 1). The AMPS-Chol-C5-MA copolymers with the Chol-C5-MA contents lower than 5 mol % are soluble in water. Solutions of these copolymers are optically clear at polymer concentrations $\leq$ 1 g/L, although solutions are slightly turbidity at concentrations $\geq$ 5 g/L. However, the copolymers with the Chol-C5-MA contents higher than 7 mol % yield strongly turbid solutions.

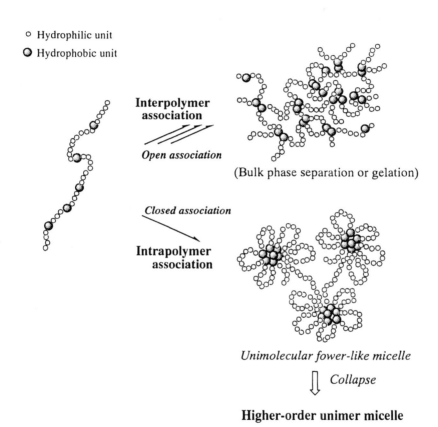

○ Hydrophilic unit

◉ Hydrophobic unit

Interpolymer association

Open association

(Bulk phase separation or gelation)

Closed association

Intrapolymer association

Unimolecular fower-like micelle

⇩ *Collapse*

Higher-order unimer micelle

Figure 1. Conceptual illustration of hydrophobic association of amphiphilic random copolymers.

The weight-average molecular weights (M_w) of all the cholesterol-containing polymers and the reference polymer [poly(A/Py)] (Chart 1), estimated by SEC measured at 70 °C using a mixed solvent of water (0.2 M phosphate buffer) and acetonitrile (50/50, v/v), are more or less the same as listed in Table 1. In contrast, M_w values for the Chol-C$_5$-MA containing polymers estimated by SLS in 0.1 M NaCl aqueous solution are much larger than those estimated by SEC, M_w markedly increasing with increasing the cholesterol content. These results indicate that interpolymer cholesterol association occurs in water, and the extent of the interpolymer association increases with increasing the cholesterol content. On the other hand, M_w values for the 0.5 and 1 mol % CholMA polymers are much lower than those of the Chol-C$_5$-MA polymers at the same cholesterol contents. The M_w values for the 0.5 and 1 mol % CholMA polymers are practically the same and slightly higher than that of the reference polymer. The apparent mean-square radii of gyration $<s^2>^{1/2}$ for the 1 mol % Chol-C$_5$-MA polymer (50.3 nm) is much larger than that of the CholMA containing polymer of the same cholesterol content (21.7 nm). Thus, there is a clear tendency that M_w and $<s^2>^{1/2}$ for the Chol-C$_5$-MA polymers are considerably larger than those for the CholMA polymers at the same cholesterol content. Therefore, it is concluded that the pentamethylene spacer makes cholesterol pendants favorably associate between polymer chains. Without the spacer, no such interpolymer association occurs when the cholesterol contents are lower than 1 mol %.

Table 1. Weight-Average Molecular Weights and Radii of Gyration for the Polymers

polymer	$M_w \times 10^4$		$<s^2>^{1/2}$ [b] (nm)
	SEC[a]	SLS[b]	
poly(A/Py)	4.8	6.2	14.1
poly(A/CholMA0.5/Py)	5.4	7.8	19.4
poly(A/CholMA1/Py)	5.5	8.4	21.7
poly(A/Chol-C$_5$-MA0.5/Py)	5.3	20.2	42.2
poly(A/Chol-C$_5$-MA1/Py)	5.5	49.3	50.3
poly(A/Chol-C$_5$-MA5/Py)	3.8	190	65.7

[a] Determination by SEC using a mixed solvent of water (0.2 M phosphate buffer) and acetonitrile (50/50, v/v) as the eluent at 70 °C.
[b] Determined by SLS in 0.1 M NaCl aqueous solution at 25 °C.

The interpolymer cholesterol association is indicated by NRET between a naphthalene-labeled polymer [poly(A/Chol-C$_5$-MA5/Np)] and a pyrene-labeled polymer [poly(A/Chol-C$_5$-MA5/Py)] (Chart 1). In Figure 2, the intensity ratios of pyrene fluorescence to naphthalene fluorescence (I_{Py}/I_{Np}) in fluorescence spectra of mixtures of the pyrene-labeled and naphthalene-labeled polymers in water are plotted as a function of the total polymer concentration. The I_{Py}/I_{Np} ratio is a measure for the quantum efficiency of NRET between the naphthalene and pyrene labels. Naphthalene can be predominantly excited at 290 nm with a slight contribution of direct excitation of pyrene, which can be corrected for the estimation of I_{Py}/I_{Np} (44–46). In a concentration range from 0.02 to 0.9 g/L, the I_{Py}/I_{Np} ratio appears to increase slightly as the polymer concentration is increased, but the ratio increases more significantly in a concentration regime > 1 g/L. This increase in I_{Py}/I_{Np} is due to a decrease in the

Chart 1

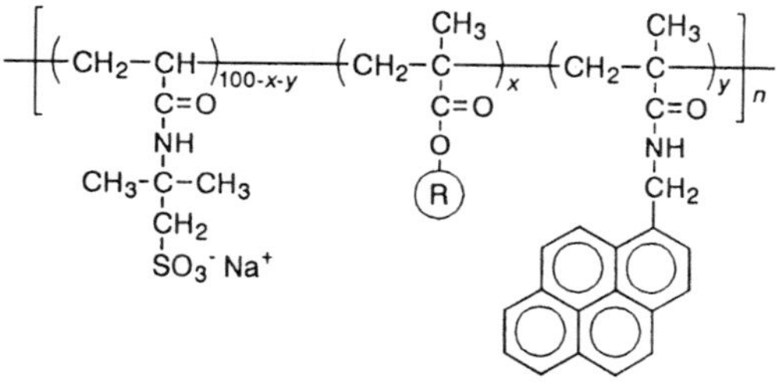

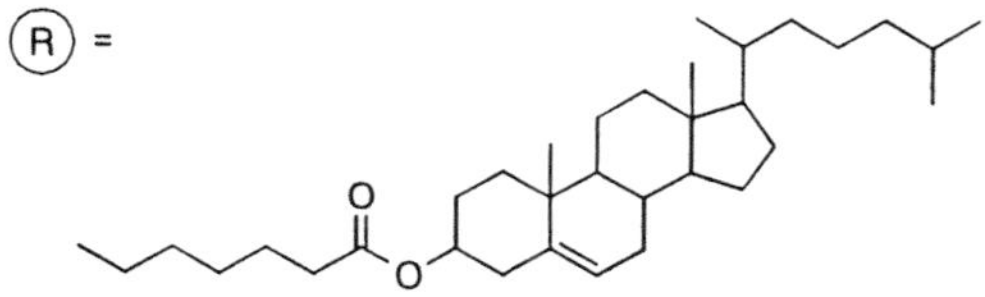

$x = 0.5$ $y = 1$ poly(A/Chol-C$_5$-MA0.5/Py)
$x = 1$ $y = 1$ poly(A/Chol-C$_5$-MA1/Py)
$x = 5$ $y = 1$ poly(A/Chol-C$_5$-MA5/Py)
$x = 7$ $y = 1$ poly(A/Chol-C$_5$-MA7/Py)
$x = 10$ $y = 1$ poly(A/Chol-C$_5$-MA10/Py)
$x = 1$ $y = 0$ poly(A/Chol-C$_5$-MA1)
$x = 5$ $y = 0$ poly(A/Chol-C$_5$-MA5)

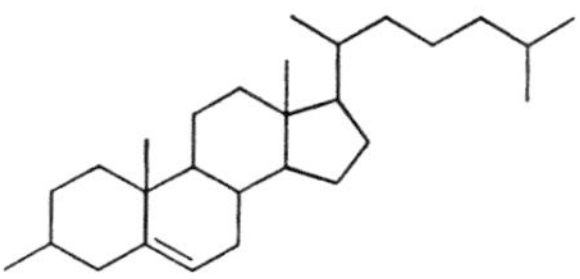

$x = 0.5$ $y = 1$ poly(A/CholMA0.5/Py)
$x = 1$ $y = 1$ poly(A/CholMA1/Py)
$x = 1$ $y = 0$ poly(A/CholMA1)
$x = 0$ $y = 1$ poly(A/Py) (reference polymer)

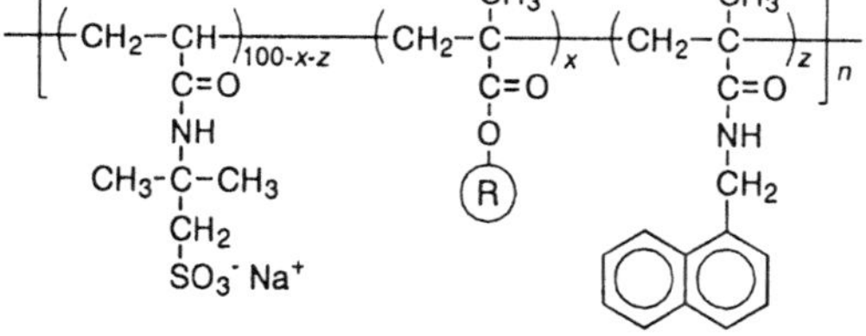

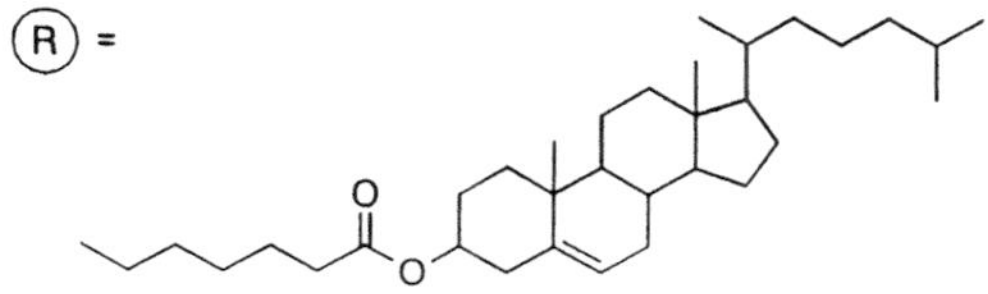

$x = 5$ $z = 1$ poly(A/Chol-C$_5$-MA5/Np)

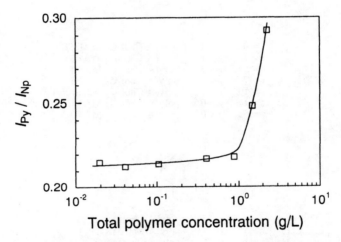

Figure 2. The I_{Py}/I_{Np} ratio as a function of the concentration of the mixture of poly(A/Chol-C$_5$-MA5/Py) and poly(A/Chol-C$_5$-MA5/Np) in pure water.

average spacing between naphthalene and pyrene labels. As the polymer concentration is increased, the fraction of naphthalene and pyrene labels that are close to each other within the Förster radius (R_0 = 2.86 nm for transfer from 1-methylnaphthalene to pyrene (47)) is increased. The results in Figure 2 are an experimental manifestation that interpolymer self-association of cholesterol groups occurs even in a very low concentration regime ($\leq$ 0.9 g/L) and that the interpolymer cholesterol association markedly increases at concentrations > 1 g/L.

The strong propensity for interpolymer association of the Chol-C$_5$-MA polymers can be clearly seen in relaxation time distributions in QELS. Figure 3 compares the relaxation time distributions for the reference polymer, CholMA polymers, and Chol-C$_5$-MA polymers at 1 g/L in 0.1 M NaCl aqueous solutions at 25 °C. The relaxation time distributions for the reference polymer and the CholMA polymers are unimodal with similar relaxation times. In contrast, the relaxation time distributions for the 0.5 and 1 mol % Chol-C$_5$-MA polymers are bimodal with fast and slow relaxation modes. Peaks for the fast mode for these polymers have relaxation times similar to those for the reference polymer and the CholMA polymers. Therefore, the peak with the fast relaxation time in these polymers can be ascribed to free (non-associated) polymers (i.e., unimers). In the relaxation time distribution for the 5 mol % Chol-C$_5$-MA polymer, the fast relaxation mode only appears as a small shoulder, indicating that most polymer chains are intermolecularly associated. If we assume that all the polymer chains are associated, the aggregation number for the 5 mol % Chol-C$_5$-MA polymer can be roughly estimated to be 50 from the molecular weights determined by SLS in 0.1 M NaCl aqueous solution and SEC in water/acetonitrile (50/50, v/v) (Table 1).

Figure 4 shows the dependencies of the ratio of the relaxation rate (Γ) and the square of the scattering vector (q^2) on q^2 at varying measuring angles in QELS for the reference polymer and the 1 and 5 mol % Chol-C$_5$-MA polymers. The Γ/q^2 ratios for the reference polymer and for the fast mode (see Figure 3) for the 1 mol % Chol-C$_5$-MA polymer are independent of q^2, indicating that the relaxation mode is practically due to a diffusive process. However, the slow mode peaks for the 1 mol % Chol-C$_5$-MA polymer and for the 5 mol % Chol-C$_5$-MA polymer are significantly angular dependent, implying that the relaxation mode is not solely due to the diffusional motion but there is a contribution of internal motions of the scatterer. Approximate values of hydrodynamic radii (R_h), calculated from the Stokes-Einstein relation along with the viscosity of water with use of approximate values of the diffusion coefficients (D) estimated from D = Γ/q^2, are 6 nm for the reference polymer and 7 nm for the fast mode for the 1 mol % Chol-C$_5$-MA polymer, whereas they are 44 nm for the slow mode for the 1 mol % Chol-C$_5$-MA polymer and 42 nm for 5 mol % Chol-C$_5$-MA polymer (estimated at 90 °). It is important to note that R_h values for the reference polymer and for the fast mode in the 1 mol % Chol-C$_5$-MA polymer are essentially the same. Thus, the fast modes observed in the Chol-C$_5$-MA polymers in Figure 3 are due to the unimer state of the polymers. Furthermore, R_h for the slow mode in the 1 mol % Chol-C$_5$-MA polymer is fairly close to that for the 5 mol % Chol-C$_5$-MA polymer estimated at an angle of 90°. In fact, the slow-mode peaks in the relaxation time distributions in Figure 3 for the Chol-C$_5$-MA polymers are quite symmetrical with similar widths and peak relaxation times independent of the Chol-C$_5$-MA content in the polymer. These observations suggest that the interpolymer association for the Chol-C$_5$-MA polymers is not an "open association" discussed in the previous section but appears to be a "closed association" where interpolymer association stops at a certain aggregation number. It is reasonable to consider that the cholesterol pendants in the Chol-C$_5$-MA polymers self-associate not only intermolecularly but also intramolecularly to form a flower-like micelle and that the concurrent intermolecular cholesterol association may link the flower-like micelles together, leading to intermolecularly-bridged flower-like micelles. A conceptual model for such micellar structure is illustrated in Figure 5.

84

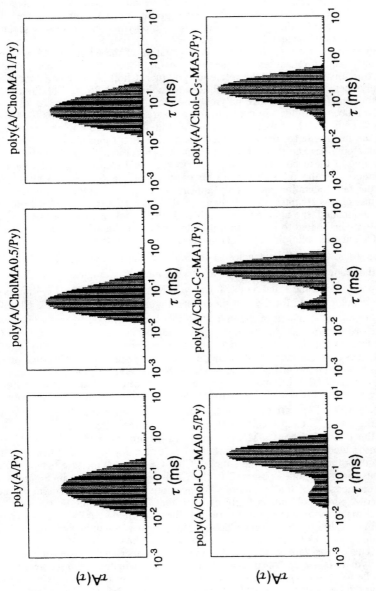

Figure 3. Relaxation time distributions for cholesterol-bearing polymers at 1 g/L in 0.1 M NaCl measured at $\theta = 90°$ at 25 °C.

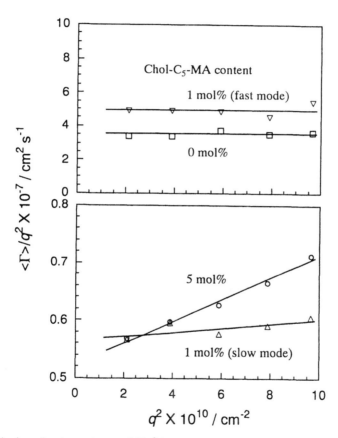

Figure 4. Angular dependence of Γ/q^2 in QELS for the Chol-C$_5$-MA polymers.

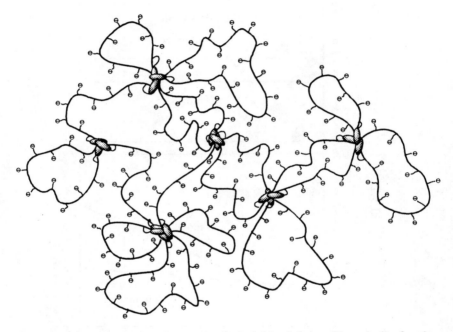

Figure 5. A model of an intermolecularly-bridged flower-like micelle for the Chol-C$_5$-MA polymer.

It may be expected that aqueous solutions of such hydrophobically "crosslinked" polymers show a shear-thinning effect. But, the viscosity of a 10 g/L aqueous solution of the 5 mol % Chol-C$_5$-MA polymer at 30 °C is independent of the shear rate up to 250 s^{-1}. This observation suggests that hydrophobic association between cholesterol pendants is tight enough not to be dissociated by mechanical shear.

Hydrophobic Association of Random Copolymers of AMPS and N-Dodecylmethacrylamide.

Hydrophobic associations in random copolymers of AMPS and N-dodecylmethacrylamide (DodMAm) were investigated by viscometry, QELS, capillary electrophoresis (CE), NMR relaxation, and various fluorescence techniques (24,48). For fluorescence studies, the polymers that are singly-labeled with pyrene or naphthalene or doubly-labeled with pyrene and naphthalene (Chart 2) were employed (48). In this section, the self-association behavior of AMPS-DodMAm copolymers in water will be discussed with a focus on intra- and interpolymer self-associations as a function of the DodMAm content (f_{Dod}) in the copolymer.

The associative behavior of the AMPS-DodMAm copolymer depends strongly on f_{Dod} (24,48). Figure 6 compares relaxation time distributions in QELS at varying concentrations for the copolymers with f_{Dod} = 2.5 and 10 mol %. The relaxation time distributions for the copolymer with f_{Dod} = 2.5 mol % are apparently unimodal. However, as the polymer concentration is increased, the distribution becomes bimodal, and the slow mode component increases with increasing the polymer concentration. In contrast, the copolymer with f_{Dod} = 10 mol % shows unimodal distributions independent of the polymer concentration. These QELS results indicate that the AMPS-DodMAm copolymers undergo interpolymer association when f_{Dod} is as low as 2.5 mol % whereas intrapolymer association becomes predominant when f_{Dod} is increased to 10 mol %. The fast mode is due to the unimer state of the polymer and the slow mode is due to intermolecular associates of the polymers. These observations in QELS are consistent with those in the viscometry. The reduced viscosity first increases as f_{Dod} is increased showing a peak at $f_{Dod} \approx$ 2.5–5 mol %, and then significantly decreases in the range 7.5 < f_{Dod} < 20 mol %.

The intermolecular association of polymer-bound dodecyl groups is evidenced by an effect of a surfactant molecule added to polymer solutions. Figure 7 compares relaxation time distributions for the copolymers with f_{Dod} = 2.5 and 10 mol % at a 10.0 g/L polymer concentration in the presence of varying concentrations of n-dodecyl hexaoxyethylene glycol monoether (C$_{12}$E$_6$). In the case of the copolymer with f_{Dod} = 2.5 mol %, the slow mode component decreases as the C$_{12}$E$_6$ concentration is increased, and the distribution becomes completely unimodal at C$_{12}$E$_6$ concentrations $\geq$ 0.5 mM. In the case of the polymer with f_{Dod} = 10 mol %, however, there is practically no effect of C$_{12}$E$_6$ added up to 5 mM. These results indicate that the interpolymer associations of dodecyl groups in the copolymer with f_{Dod} = 2.5 mol % are dissociated by interactions with C$_{12}$E$_6$. The hydrophobic association of the polymer-bound dodecyl groups with C$_{12}$E$_6$ is more favorable than their self-association (49).

When f_{Dod} is increased beyond 10 mol %, hydrophobic microdomains of polymer-bound dodecyl groups are progressively formed by intrapolymer hydrophobe associations. The formation of the hydrophobic microdomains leads to a considerable chain collapse. The chain collapse occurs most significantly in an f_{Dod} range of 10–30 mol % in 0.1 M NaCl aqueous solutions. These hydrophobic microdomains are surrounded by loops of charged segments, and thus one may conjecture a flower-like unimolecular micelle (unimer micelle) where a number of hydrophobic microdomains are joined together in clusters on a polymer chain, as conceptually illustrated in Figure 8.

88

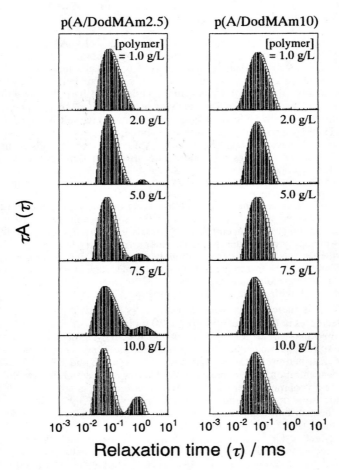

Figure 6. Relaxation time distributions for the AMPS-DodMAm copolymers with f_{Dod} = 2.5 and 10 mol % at varying concentrations in 0.1 M NaCl at θ = 90° at 25 °C.

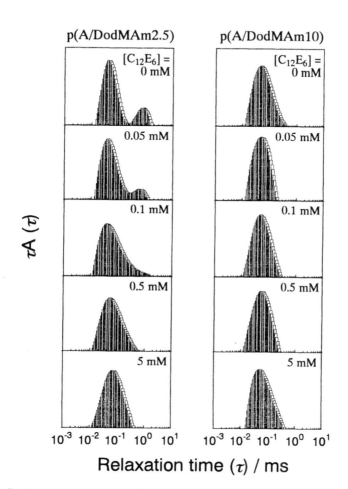

Figure 7. Relaxation time distributions for the AMPS-DodMAm copolymers with f_{Dod} = 2.5 and 10 mol % at 10.0 g/L in 0.1 M NaCl at varying concentrations of added $C_{12}E_6$ at θ = 90° at 25 °C.

Hydrophobe < 10 mol%

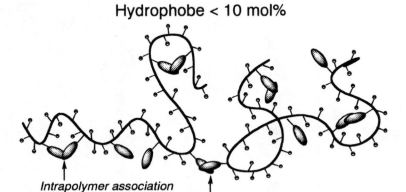

Intrapolymer association

Interpolymer association

Hydrophobe ≈ 10 - 30 mol%

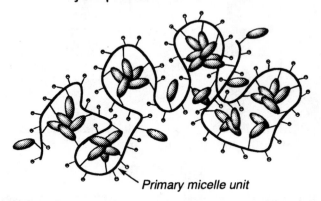

Primary micelle unit

Figure 8.   Conceptual models for inter- and intrapolymer hydrophobic associations.

The intensity ratio of the third to first peaks in the vibrational fine structures in pyrene fluorescence spectra (I_3/I_1) is often used as an indicator for the microenvironmental polarity about pyrene. In general, the I_3/I_1 ratio is larger in less polar microenvironments (50). The I_3/I_1 ratio of the pyrene-labeled polymers (Chart 2) with varying f_{Dod} increases in the region $10 < f_{Dod} < 30$ mol %, reaching a maximum at $f_{Dod} \approx 30$ mol % (Figure 9) (48), indicating that hydrophobic microdomains begin to form at $f_{Dod} \approx 10$ mol % and the pyrene labels are entrapped in the hydrophobic microdomains. With an increase in f_{Dod}, the extent of the label incorporation increases, reaching a maximum extent at $f_{Dod} \approx 30$ mol%.

Intrapolymer NRET between naphthalene and pyrene, both labeled on the same polymer chain, provides information about changes in the microscopic chain dimension (46). The intensity of pyrene fluorescence emitted from the doubly-labeled polymers (Chart 2) upon excitation of naphthalene at 290 nm increases significantly with increasing f_{Dod} (Figure 10), arising from intrapolymer NRET from photoexcited naphthalene to pyrene labels (48). This is due to a decrease in the average separation between the naphthalene and pyrene labels within the same polymer chain (i.e., with increasing f_{Dod}, increasing fractions of naphthalene and pyrene labels come close to each other within the Förster radius ($R_0 = 2.86$ nm (47)). This decrease in the donor-acceptor distance results from an increase in the extent of the intrapolymer association of dodecyl groups with increasing f_{Dod}. The I_{Py}/I_{Np} ratio slightly increases at $f_{Dod} \leq 10$ mol% and more significantly increases in the range $10 < f_{Dod} < 40$ mol %, reaching a plateau at $f_{Dod} \approx 40$ mol% (Figure 10), indicate that the compaction of the polymer size occurs over a range of $10 < f_{Dod} < 30$ mol % and saturates at $f_{Dod} \approx 40$ mol%. These results are practically consistent with those of the I_3/I_1 ratio (Figure 9). NRET reflects the conformational compactness, whereas the I_3/I_1 ratio reflects the micropolarity around fluorescence labels. Therefore, as f_{Dod} is increased, the polymer compaction occurs parallel to an increase in the micropolarity about the fluorescence labels.

As f_{Dod} is further increased to 50 mol % or higher, the polymers form compact unimer micelles where polymer chains are highly collapsed, arising from secondary association between the surfaces of hydrophobic microdomains within the same polymer chain. The flower-like unimer micelles formed at $f_{Dod} \approx 10$–40 mol % may be viewed as a secondary structure of the polymer. At $f_{Dod} > 40$ mol %, however, the secondary structure is folded into a tertiary structure comprised of an aggregated cluster of the hydrophobic microdomains (Figure 11).

The collapse of the secondary structure into the tertiary structure can be observed by capillary electrophoresis (24). Electropherograms for the AMPS-DodMAm copolymers with varying f_{Dod} are shown in Figure 12. The migration time decreases as f_{Dod} is increased because the electric charge decreases accordingly. In general, the migration time is determined by the total charge and hydrodynamic volume of a polymer. Therefore, the band width of an electropherogram for a copolymer of ionic and nonionic comonomers is an indication of the distribution of copolymer compositions, molecular weights, or hydrodynamic volumes of the copolymer. Because AMPS and DodMAm undergo "ideal" copolymerization with the monomer reactivity ratios being essentially unity (51), it is unlikely that there is a difference in the compositional distribution and molecular weight distribution among the copolymers with different f_{Dod}. Therefore, the band width of the electropherogram should be an indication of the distribution of the hydrodynamic volume. As can be seen in Figure 12, the band becomes broader as f_{Dod} is increased up to 30 mol %, implying that the distribution of the hydrodynamic volumes (i.e., the conformational distributions) for the secondary structure of these copolymers gets broader with increasing f_{Dod} up to 30 mol %. When f_{Dod} is increased to 40 mol %, however, the band abruptly becomes

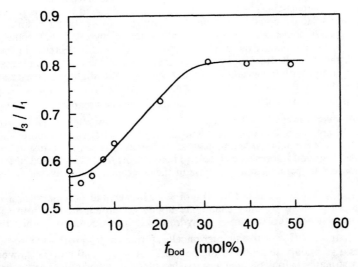

Figure 9. The I_3/I_1 ratio for the pyrene-labeled AMPS-DodMAm copolymers in 0.1 M NaCl as a function of f_{Dod}.

Chart 2

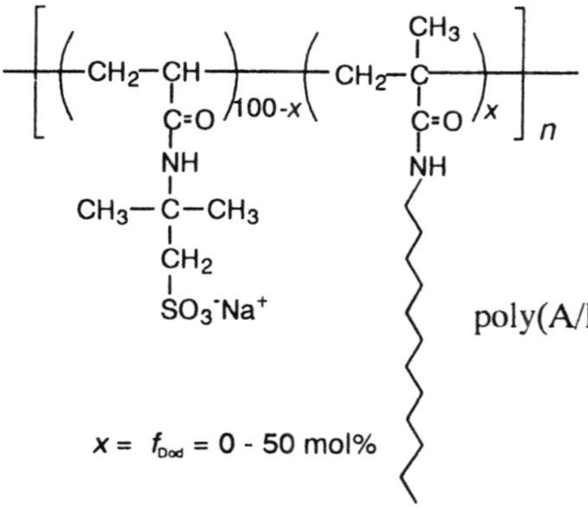

poly(A/DodMAm*x*)

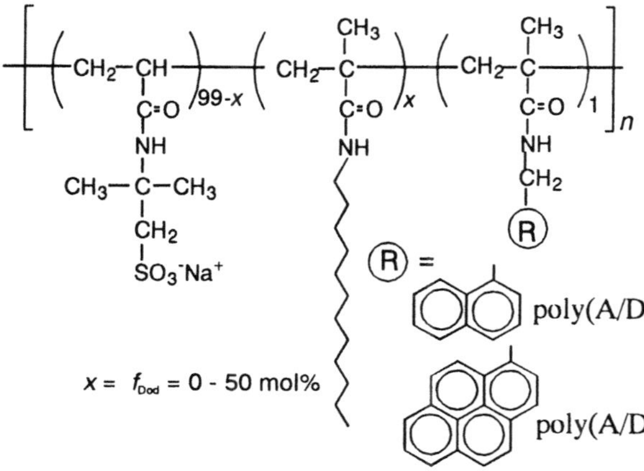

$x = f_{Dod} = 0 - 50$ mol%

$x = f_{Dod} = 0 - 50$ mol%

$(R) =$

poly(A/DodMAm*x*/Np)

poly(A/DodMAm*x*/Py)

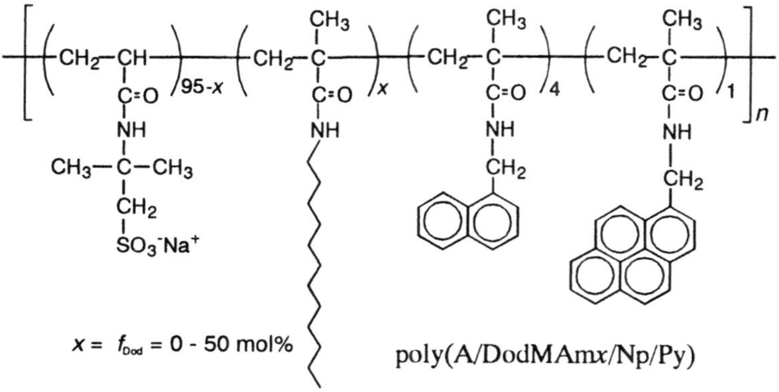

$x = f_{Dod} = 0 - 50$ mol%

poly(A/DodMAm*x*/Np/Py)

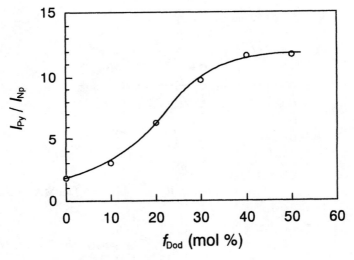

Figure 10. The I_{Py}/I_{Np} ratio for the doubly-labeled polymers in 0.1 M NaCl as a function of f_{Dod}.

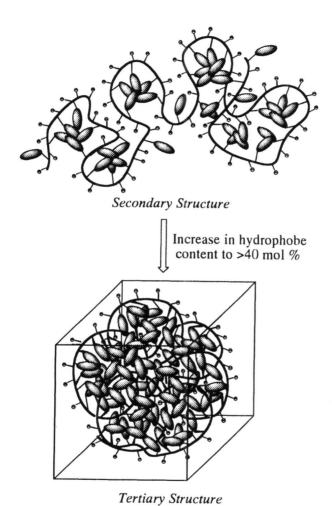

Secondary Structure

Increase in hydrophobe
content to >40 mol %

Tertiary Structure

Figure 11. Conceptual illustration of the collapse of the secondary structure into the tertiary structure with an increase in the hydrophobe content.

very sharp, an indication of the collapse of the secondary structure into the tertiary structure of the unimer micelle. These observations in the capillary electrophoresis are consistent with the observations in QELS. QELS results indicate that the apparent hydrodynamic radius markedly decreases with increasing f_{Dod} in the region $10 < f_{Dod} < 30$ mol %, and the relaxation time distribution becomes very sharp as f_{Dod} is increased to 40 mol %.

Dynamic Properties of Higher-Order Unimer Micelles.

Not only the copolymers of AMPS and DodMAm but also the copolymers of AMPS and a methacrylamide N-substituted with cyclododecyl, adamantyl, or naphthylmethyl groups (Chart 3) form either a unimolecular flower-like or higher-order micelle (i.e., tertiary structure) depending on the hydrophobe content in the copolymer. An important feature for these copolymers is that such unimer micelles are formed even at high polymer concentrations in water. Some of the requisite structures for polymers to form unimer micelles are; (i) the hydrophobes should be sufficiently bulky, (ii) the sequence distribution of hydrophobic and electrolyte monomer units should be random, and (iii) the hydrophobes should be linked to the polymer main chain via amide spacer bonding (22).

A characteristic feature for higher-order unimer micelles is that the size is extremely small for its molecular mass. For example, a unimer micelle formed from a 1 : 1 copolymer of AMPS and cyclododecylmethacrylamide with a weight-average molecular weight of 5.1×10^5 shows a hydrodynamic diameter of 11 nm (19). A 10 wt % aqueous solution of this unimer micelle shows a scattering peak in small-angle X-ray scattering (SAXS) corresponding to a spacing of 11 nm which agrees with the hydrodynamic diameter, arising from a macro-lattice formed by globular unimer micelles closely packed with a spacing corresponding to the diameter of the unimer micelle in a concentrated aqueous solution (19). These observations indicate that the higher-order unimer micelles remain as such even at high concentrations.

Among the hydrophobes, there a significant difference in the extent of the "protection" of pyrene labels in the copolymers. The micropolarity of the hydrophobic microdomains probed by the I_3/I_1 ratio of pyrene fluorescence decreases in the order of adamantyl > dodecyl > cyclododecyl domains (51). Furthermore, the protection of pyrene fluorescence from quenching by thallium cations in the bulk aqueous phase is much more effective in cyclododecyl domains than in dodecyl domains (19).

The extent of the restriction on the mobility of an azobenzene moiety covalently incorporated into higher-order unimer micelles, monitored by photoisomerization and thermal back isomerization rates, is much more pronounced in cyclododecyl domains than in dodecyl domains, which can be related to a difference in the mobility of these aliphatic hydrophobes in the microdomains (52). Cyclododecane is a rigid molecule with much less conformational freedom than dodecane because of its cyclic structure. The latter is a flexible chain that can adopt a variety of conformations. Therefore, cyclododecyl groups may be more tightly associated than dodecyl groups and form a more rigid hydrophobic microdomain.

Functionalization of Higher-Order Unimer Micelles.

Hydrophobic functional-molecules can be incorporated into the higher-order unimer micelles either via chemical bonding of the functional molecules to amphiphilic polymers or by physically solubilizing the molecules into the unimer micelle. For example, when an AMPS-DodMAm copolymer and a hydrophobic small-molecule are dissolved together in ethanol, and the ethanol solution is poured into excess water, then the hydrophobic molecules can be entrapped in a hydrophobic microdomain during the process of the hydrophobic self-association of polymer-bound dodecyl groups. However, it is practically difficult to precisely control the number of the small

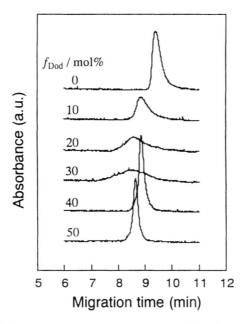

Figure 12. Electropherograms for the AMPS-DodMAm copolymers with varying f_{Dod} measured in a borate buffer (pH = 9.03, ionic strength 0.05).

Chart 3

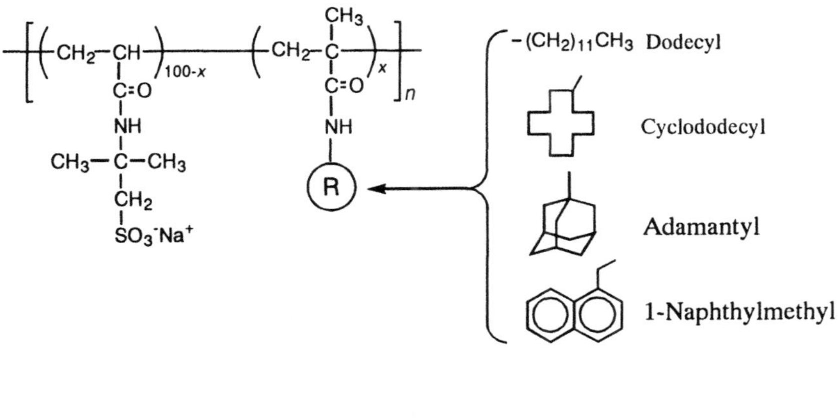

molecules incorporated in a unimer micelle by this simple method. One can easily incorporate a predetermined number of functional molecules into a unimer micelle by a terpolymerization technique using a functional molecule substituted with a polymerizable group. The functional groups thus incorporated can be embedded in the hydrophobic microdomains in the unimer micelle, leading to the "compartmentalization" or "nanoencapsulation" of functional groups in the unimer micelle (19,21,22,51,52).

Over the last decade, there has been increasing interest in photophysics and photochemistry in constraining space. If photoactive chromophores are compartmentalized in the hydrophobic microdomains of the unimer micelle, each chromophore molecule will be completely isolated from others in highly constraining nonpolar microenvironments and protected from the aqueous phase. Arising from these unusual microenvironmental effects, such chromophore-functionalized unimer micelles, unlike the conventional molecular assembly systems, may induce a large modification of the photophysical and photochemical behavior of the compartmentalized chromophores. This section will briefly present some examples of such unusual photophysical and photochemical behavior of chromophores encapsulated in the higher-order unimer micelle.

Anomalous Photochemistry of a Porphyrin Moiety Encapsulated in Unimer Micelles. In a variety of biological systems, metalloporphyrins are embedded in protein environments (53–55) that play a critical role in controlling reactivities and biological functions of the metalloporphyrins. In the reaction center of natural photosynthetic systems, for instance, chlorophylls are assembled in protein matrices with a specific alignment such that the quantum efficiencies of photoexcited energy transfer and photoinduced electron transfer are maximized. Studies of the effects of generic microenvironments on the photophysical and photochemical properties of metalloporphyrins in model systems have received considerable interest from chemists because such studies may provide not only a mechanistic insight into the biological photosynthesis but also a guideline for designing of artificial photosystems for photon energy conversion into electronic energy or chemical potential.

Zinc(II) tetraphenylporphyrin (ZnTPP) can be covalently incorporated into the unimer micelle by terpolymerization using a ZnTPP-substituted acrylamide monomer Zinc(II) 5-(4-acrylamidophenyl)-10,15,20-triphenylporphynate (ZnAATPP) (Chart 4) (56). The ZnTPP units compartmentalized within the higher-order unimer micelle are protected from the bulk aqueous phase. Because the amount of the ZnTPP units incorporated in the terpolymers are very small (0.1–0.2 mol %), each ZnTPP unit is independently compartmentalized in the hydrophobic microdomain and is prohibited from encountering others. In the reference copolymer (Chart 4), on the other hand, the ZnTPP moieties are exposed to the aqueous phase.

The photophysical properties of the ZnTPP chromophores compartmentalized in the unimer micelle are greatly modified as compared to those in homogeneous solutions. The most remarkable of all is that the lifetime of the triplet excited state of ZnTPP chromophores in the unimer micelle becomes extraordinarily long in aqueous solution at ordinary or higher temperature (57). This is particularly true in cyclododecyl domains formed from the terpolymer poly(A/CD/ZnTPP). The decay profiles of triplet-triplet absorption for the compartmentalized ZnTPP systems are fitted with double-exponential functions, from which average lifetimes are estimated. The triplet lifetime of ZnTPP in the reference copolymer is more or less 2 ms, which is about the same as the lifetime of small molecular-weight ZnTPP in organic solvents. In contrast, the average lifetime of ZnTPP triplets in cyclododecyl domains is 58 ms, whereas it is 17 ms in dodecyl domains. This extraordinarily long triplet lifetime in cyclododecyl domains is attributable to the rigidity of the microdomain to which ZnTPP is tightly "pinned down" (57).

Because of the unusually long lifetime of the ZnTPP triplets in cyclododecyl

Chart 4

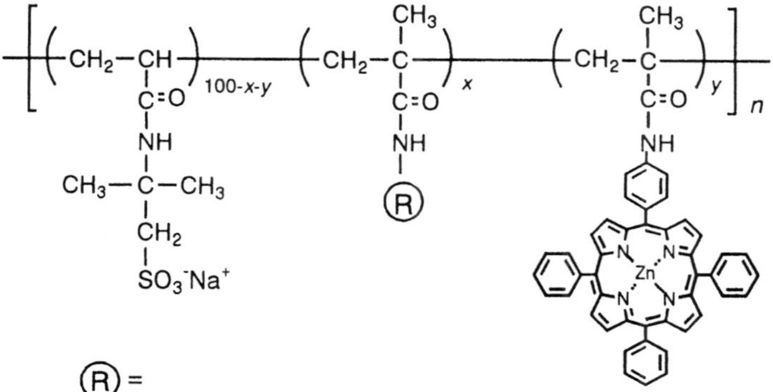

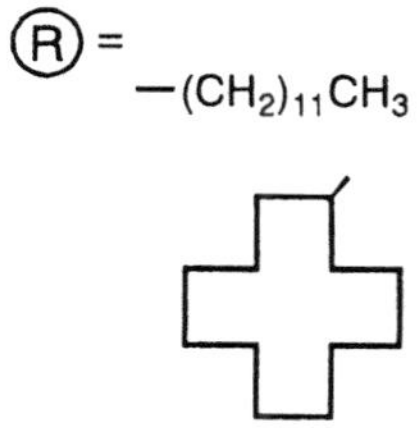

$\text{(R)} = $ $-(CH_2)_{11}CH_3$ poly(A/Dod/ZnTPP)
$x = 61$ mol %, $y = 0.19$ mol %

poly(A/CD/ZnTPP)
$x = 60$ mol %, $y = 0.084$ mol %

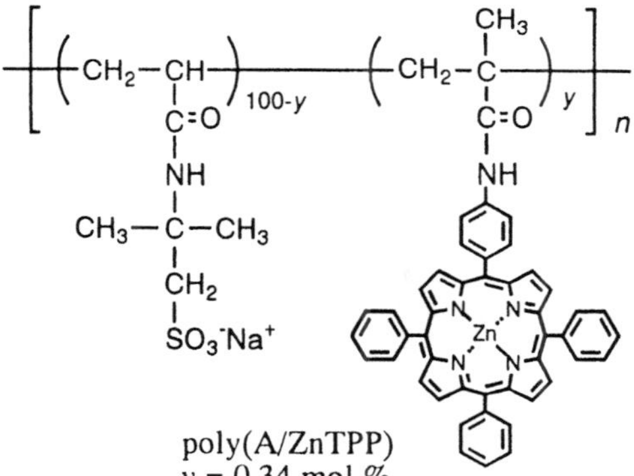

poly(A/ZnTPP)
$y = 0.34$ mol %
Reference copolymer

domains, ZnTPP in the unimer micelle emits intensive phosphorescence and "E-type" delayed fluorescence in aqueous solution at ordinary or higher temperatures (58), which is a unusual phenomenon for Zn-porphyrins. The reference copolymer does not emit such delayed fluorescence and phosphorescence in aqueous solution at room temperature. The delayed emissions show a characteristic temperature dependence; with increasing temperature, the delayed fluorescence increases whereas the phosphorescence decreases. Analysis of the temperature dependence data indicates that the delayed fluorescence is due to the thermal activation of the triplet excited state back to the singlet excited state. The ZnTPP chromophores encapsulated in dodecyl microdomains also show phosphorescence and delayed fluorescence in aqueous solution at ordinary temperatures, although the intensities are much less than those in cyclododecyl microdomains.

The extremely long-lived ZnTPP triplets in cyclododecyl domains in the unimer micelle facilitate electron transfer reaction to electron acceptors. Phenylmethylphenacylsulfonium p-toluenesulfonate (PMPS), an onium salt, is known as a self-destructive electron acceptor which decomposes rapidly into phenylmethylsulfide and a phenacyl radical upon accepting an electron (59). When PMPS is added to an aqueous solution of ZnTPP-functionalized unimer micelles, the electron acceptor is localized on the anionic surface of the unimer micelle. The electron transfer from the ZnTPP triplet to PMPS produces a ZnTPP cation radical (ZnTPP$^{+}\bullet$) and a phenacyl radical as transient species (Scheme 1), but reaction between these two transient species, which usually follows the photoinduced forward electron transfer in homogeneous solutions, is prevented in the unimer micelle system because the ZnTPP$^{+}\bullet$ and phenacyl radical species are separated. Thus, the porphyrin cation radicals can be accumulated in the unimer micelle.

Figure 13 shows the decays of ESR signal due to ZnTPP$^{+}\bullet$ produced by irradiation of visible light for 1 min (60). Significant amounts of the porphyrin cation radicals are accumulated in the unimer micelle systems. A small ESR signal for the reference polymer system decays almost instantly when the light is turned off. This decay of the ESR signal is probably due to rereduction of ZnTPP$^{+}\bullet$ by an electron from water. The decay of the ESR signal in the cyclododecyl system is much slower than that in the dodecyl system. In the cyclododecyl unimer micelle, the ZnTPP$^{+}\bullet$ species are much better protected from the aqueous phase than those in the dodecyl unimer micelle, leading to a much longer lifetime of the porphyrin cation radical in the former.

Anomalous Emission Properties of a C_{60} Moiety Encapsulated in Unimer Micelles. To encapsulate a C_{60} moiety into the higher-order unimer micelle, a terpolymer of AMPS (49.5 mol %), N-cyclododecylmethacrylamide (49.5 mol %), and a methacrylate carrying a C_{60} moiety (0.1 mol %) was synthesized (Chart 5) (61). The weight-average molecular weight and the apparent hydrophobic diameter of the C_{60}-carrying polymer are 1×10^5 (SLS) and 13 nm (QELS) in 0.1 M NaCl, respectively. Since the C_{60} moiety is highly hydrophobic, it would be incorporated in the hydrophobic microdomain in the higher-order unimer micelle in aqueous solution. Absorption spectra for the C_{60} moiety in the unimer micelle shows a broad absorption over the range of 240–700 nm with a peak around 420 nm, which is characteristic of C_{60} chromophores. Figure 14 compares fluorescence spectra for the C_{60} moiety in the unimer micelle in water and for the reference C_{60} compound (Chart 5) in tetrahydrofuran. The fluorescence maximum for the C_{60} moiety in the unimer micelle is blue-shifted by 150 nm comparing to that of the reference compound. It is reported that the fluorescence of C_{60} chromophores shows a tendency for a blue shift when incorporated into polymers (62,63). The shortest fluorescence maximum reported so far for polymer-bound C_{60} chromophores is 600 nm. The much larger blue shift for the C_{60} moiety in the unimer micelle, showing a maximum at 560 nm, may be

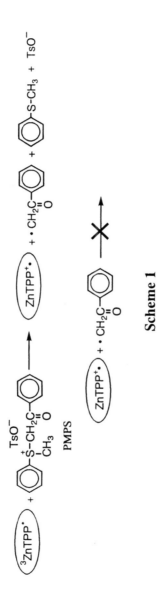

Scheme 1

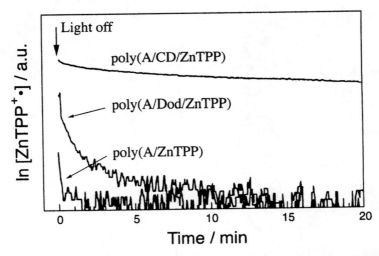

Figure 13. Decay profiles for the ESR absorption after irradiation with visible light for 1 min in the presence of PMPS in aqueous solutions.

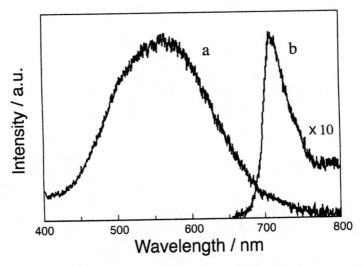

Figure 14. Fluorescence spectra for the C_{60} moiety in the higher-order unimer micelle in water and for the reference C_{60} compound in tetrahydrofuran.

Chart 5

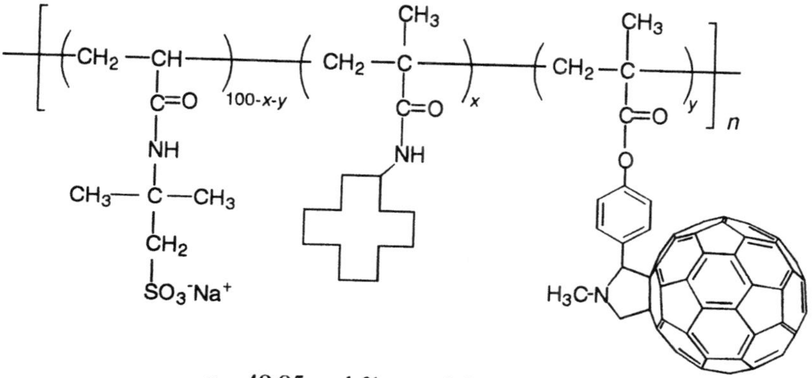

$x = 49.95$ mol %, $y = 0.1$ mol %

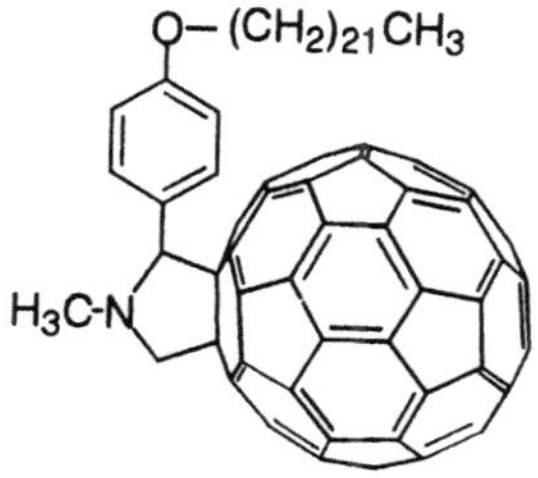

Reference compound

attributed to a highly constraining hydrophobic microenvironment in the unimer micelle.

Concluding Remarks

The higher-order unimer micelles somewhat resemble globular proteins in that (i) the unimer micelles are a spherical object with a tertiary structure rich in charged groups on the surface and hydrophobic groups in the interior, (ii) the unimer micelles stay as such even at very high concentrations in water, and (iii) the tertiary structure of the unimer micelle can be "denatured" by organic solvents or a much smaller amount of surfactants. The size of the unimer micelle can be controlled by the polymer molecular weight. One of the applications of such unimer micelles includes nanoencapsulation of small molecules into the micelle either by covalent incorporation or by direct dissolution, which may provide various possibilities of practical applications. Furthermore, the unimer micelle may be used as a charged nanoparticle having a hydrophobic region on the surface, which may provide an opportunity to investigate interactions of the unimer micelle with other molecules and colloids including proteins and enzymes (64).

Literature Cited

1. McCormick, C. L.; Bock, J.; Schulz, D. N. *Encyclopedia of Polymer Science and Engineering;* John Wiley: New York, NY, 1989; Vol. 17, pp 730.
2. Bock, J.; Varadaraj, R.; Schulz, D. N.; Maurer, J. J. In *Macromolecular Complexes in Chemistry and Biology;* Dubin, P.; Bock, J.; Davies, R. M.; Schulz, D. N.; Thies, C., Eds.; Springer-Verlag: Berlin and Heidelberg, 1994, pp 33–50.
3. Valint Jr., P. L.; Bock, J.; Schulz, D. N. In *Polymers in Aqueous Media: Performance through Association;* Glass, J. E., Ed.; Advances in Chemistry Series 223; American Chemical Society: Washington D. C., 1989, pp 399.
4. McCormick, C. L.; Nonaka, T.; Johnson, C. B. *Polymer* **1988**, *29*, 731.
5. Ezzell, S. A.; McCormick, C. L. *Macromolecules* **1992**, *25*, 1881.
6. Ezzell, S. A.; Hoyle, C. E.; Creed, D. McCormick, C. L. *Macromolecules* **1992**, *25*, 1887.
7. Kotin, L.; Nagasawa, M. *J. Chem. Phys.* **1962**,*36*, 873.
8. Nagasawa, M.; Murase, T.; Kondo, K. *J. Phys. Chem.* **1965**, *69*, 4005.
9. Joyce, D. E.; Kurucse, T. (1981), *Polymer* **1981**, *22*, 415.
10. Morcellet-Sauvage, J.; Morcellet, M.; Loucheux, C. (1981), *Makromol. Chem.* **1981**, *182*, 949.
11. Chang, Y.; McCormick, C. L. *Macromolecules* **1993**, *26*, 6121.
12. McCormick, C. L.; Chang, Y. *Macromolecules* **1994**, *27*, 2151.
13. Kramer, M. C.; Welch, C. G.; Steger, J. R.; McCormick, C. L. *Macromolecules* **1995**, *28*, 5248.
14. Hu, Y.; Kramer, M. C.; Boudreaux, C. J.; McCormick, C. L. *Macromolecules* **1995**, *28*, 7100.
15. Branham, K. D.; Snowden, H. S.; McCormick, C. L. *Macromolecules* **1996**, *29*, 254.
16. Kramer, M. C.; Steger, J. R.; Hu, Y.; McCormick, C. L. *Macromolecules* **1996**, *29*, 1992.
17. Hu. Y.; Smith, G. L; Richardson, M. F.; McCormick, C. L. *Macromolecules* **1997**, *30*, 3526.
18. Hu. Y.; Armentrout, R. S.; McCormick, C. L. *Macromolecules* **1997**, *30*, 3538.
19. Morishima, Y.; Nomura, S.; Ikeda, T.; Seki, M.; Kamachi, M. *Macromolecules* **1995**, *28*, 2874.

20. Yusa, S.; M. Kamachi, M.; Morishima, Y. *Langmuir* **1998**, *14*, 6059.
21. Morishima, Y. *Trends Polym. Sci.* **1994**, *2*, 31.
22. Morishima, Y. In *Solvents and Self-Organization of Polymers*, Webber, S. E.; Tuzar, D.; Munk, P., Eds., Kluwer Academic Publishers, Dordrecht, The Netherlands, **1996**, P. 331.
23. Noda, T.; Hashidzume, A.; Morishima, Y. to be published.
24. Yamamoto, Hashidzume, A. Morishima, Y. to be published.
25. See for example: Webber, S. E. *J. Phys. Chem. B* **1998**, *102*, 2618.
26. Tanford, C *The Hydrophobic Effects*, 2nd ed., Wiley, New York, 1980.
27. Elias, H. G. *J. Macromol. Sci.*, Part A, **1973**, *7*, 601.
28. Tuzar Z.; and Kratochvil, P. *Surface and Colloid Acience;* Matijevic, E., Ed.; Plenum Press: New York, 1993.
29. Zhou, Z.; Chu, B. *Macromolecules* **1987**, *20*, 3089.
30. Wanka, G.; Hoffmann, H.; Ulbricht, W. *Colloid Polym. Sci.* **1990**, *268*, 101.
31. Brown, W.; Schillén, K.; Almgren, M.; Hvidt, S.; Bahadur, P. *J. Phys. Chem.* **1991**, *95*, 1850.
32. Glatter, O.; Scherf, G.; Schillén, K.; Brown, W. *Macromolecules* **1994**, *27*, 6046.
33. Wu, G.; Chu, B.; Schneider, D. K. *J. Phys. Chem.* **1995**, *99*, 5094.
34. Ramireddy, C; Tuzar, Z.; Procházka, K.; Webber, S. E.; Munk, P. *Macromolecules* **1992**, *25*, 2541.
35. Chan, J.; Fox, S.; Kiserow, D.; Ramireddy, C.; Munk, P.; Webber, S. E. *Macromolecules* **1993**, *26*, 7016.
36. Qin, A.; Tian, M.; Ramireddy, C.; Webber, D. E.; Munk, P. *Macromolecules* **1994**, *27*, 120.
37. Zhang, L.; Eisenberg, A. *Science* **1995**, *268*, 1728.
38. Halperin, A. *Macromolecules* **1991**, *24*, 1418.
39. Semenov, A. N.; Joanny, J.-F.; Khokhlov, A. R. *Macromolecules* **1995**, *28*, 1066.
40. Demel, R. A.; de Kruyff, B. *Biochem. Biophys. Acta* **1976**, *457*, 109.
41. Ringsdorf, H.; Schlarb, B.; Venzmer, J. *Angew. Chem. Int. Ed. Engl.* **1988**, *27*, 113.
42. Akiyoshi, K.; Deguchi, S.; Moriguchi, N.; Yamaguchi, S.; Sunamoto, J. *Macromolecules* **1993**, *26*, 3062.
43. Nishikawa, T.; Akiyoshi, K.; Sunamoto, J. *Macromolecules* **1994**, *27*, 7654.
44. Webber, S. E. *Chem. Rev.* **1990**, *90*, 1469.
45. Winnik, F. M. *Polymer* **1990**, *31*, 2125.
46. Ringsdorf, H.; Simon, J.; Winnik, F. M. *Macromolecules* **1992**, *25*, 5353 and 7306.
47. Berlman, I. B. *Energy transfer parameters of Aromatic Compounds*; Academic Press: New York, 1973.
48. Yamamoto, H.; Mizusaki, M.; Yoda, K.; Morishima, Y. *Macromolecules* **1998**, *31*, 3588.
49. Hashidzume, A.; Mizusaki, M.; Yoda, K.; Morishima, Y. *Langmuir* to be published.
50. Kalyanasundaram, K.; Thomas, J. K. *J. Am. Chem. Soc.* **1977**, *99*, 2039.
51. Morishima, Y.; Tominaga, Y.; Kamachi, M.; Okada, T.; Hirata, Y.; Mataga, N. *J. Phys. Chem.* **1991**, *95*, 6027.
52. Morishima, Y.; Tsuji, M.; Seki, M.; Kamachi, M. *Macromolecules* **1993**, *26*, 3299.
53. Hewson, W. D.; Hager, L. P. In *The Porphyrins;* Dolphin, D. Ed.; Academic Press: New York, NY, 1979, Vol. 7, Part B; pp 295.
54. Ferguson-Miller, S.; Brautigan, D. L.; Margoliash, E. In *The Porphyrins;* Dolphin, D. Ed.; Academic Press: New York, NY, 1979, Vol. 7, Part B; pp 149.

55. Gibson, Q. H. In *The Porphyrins;* Dolphin, D. Ed.; Academic Press: New York, NY, 1979, Vol. 7, Part C; pp 153.
56. Aota, H.; Morishima, Y.; Kamachi, M. *Photochem. Photobiol.* **1993**, *57*, 989.
57. Morishima, Y.; Saegusa, K.; Kamachi, M. *J. Phys. Chem.* **1995**, *99*, 4512.
58. Morishima, Y.; Saegusa, K.; Kamachi, M. *Macromolecules*, **1995**, *28*, 1203.
59. DeVoe, R. J.; Sahyun, M. R. V.; Serpone, N.; Sharma, D. K. *Can. J. Chem.* **1987**, *65*, 2342.
60. Morishima, Y.; Aota, H.; Saegusa, K.; Kamachi, M. *Macromolecules*, **1995**, *29*, 1203.
61. Higashida, S.; Nishiyama, K.; Yusa, S.; Morishima, Y.; Janot, J.-M.; Seta, P.; Imahori, H.; Kaneda, T.; Sakata, Y. *Chem. Lett.* **1998**, 381.
62. Bunker, C. E.; Lawson, G. E.; Sun, Y.-P. *Macromolecules* **1995**, *28*, 3744.
63. Sun, Y.-P.; Bunker, C. E.; Liu, B. *Chem. Phys. Lett.* **1997**, *272*, 25.
64. Sato, T.; Kamachi, M.; Mizusaki, M.; Yoda, K.; Morishima, Y. *Macromoleules* **1998**, *31*, 6871.

Chapter 8

Dendritic Macromolecules: Hype or Unique Specialty Materials

Craig J. Hawker and Marcelo Piotti

Center for Polymeric Interfaces and Macromolecular Assemblies,
IBM Almaden Research Center, 650 Harry Road, San Jose, CA 95120–6099

The ultimate success of dendritic macromolecules as a new class of specialty polymer will depend on these novel materials possessing either new and/or improved physical, mechanical, or chemical properties when compared to standard linear polymers. In this article, the difference between dendrimers and their linear or hyperbranched analogs will be examined and the effect of these different architectures on physical properties discussed.

The field of highly branched, 3-dimensional macromolecules has progressed at a torrid pace in the last 10 years with three commercial materials becoming available and a number of products being developed based on the use of these novel polymers. As a separate class of polymeric material, distinct from traditional linear, graft, and star polymers, highly branched 3-dimensional macromolecules can be divided into 2 families of related materials, near-perfectly branched and monodisperse dendrimers and polydisperse, defect-containing hyperbranched macromolecules. While related, the synthesis and structure of these materials are distinct which leads to different potential applications for these novel specialty polymers.

Interestingly, one of the most paradoxical aspects of research in the field of dendritic macromolecules is the assumption that dendrimers are inherently different from linear polymers and it is this difference that results in new and/or improved properties. While this is most probably true for larger dendrimers, it may not be true for smaller dendrimers or for some unusual classes of dendritic macromolecules. Since the field of dendritic macromolecules is rapidly expanding and becoming a major area of research it is prudent to examine in much greater detail this assumption (1). Unfortunately there have been few systematic studies comparing dendrimers with other macromolecular architectures, such as linear polymers, and any comparisons that have been made were performed with polydisperse samples of significantly different repeat unit structure. For example, the unique melt viscosity behavior of polyether dendrimers was compared with linear polystyrene and not with monodisperse linear analogs containing the same number of polyether repeat units based on 3,5-dihydroxybenzyl alcohol (2). Because of this, important issues, such as i) effect of the numerous chain end functional groups, ii) the effect of branching and, iii) the development of a well defined three-dimensional architecture cannot be addressed and the underlying reason for these inherent differences remains a mystery.

In this article, we will discuss the different approaches and attempts that have been made to answer these fundamentally important issues, not only for dendrimers but also for highly branched macromolecular architectures in general. Interestingly, all of these approaches address an intriguing aspect of polymer chemistry that has received only minor attention up till now, that is the concept of macromolecular isomers. While the concept of structural isomers is well know in small molecule chemistry, the application of similar ideas to macromolecular chemistry has not been possible until recently. As will be shown below, the development of new synthetic approaches allows the preparation of well defined, monodisperse macromolecules with significantly different branching patterns and hence 3-dimensional structure.

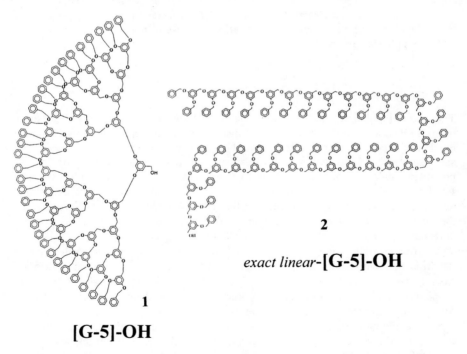

2

exact linear-[**G-5**]**-OH**

1

[**G-5**]**-OH**

Figure 1. Comparative structures of the fifth generation dendritic alcohol, **1**, and its exact linear analog, **2**.

Discussion

Construction of Linear Analogs One of the most widely studied families of dendritic macromolecules are the polyether dendrimers based on 3,5-dihydroxybenzyl alcohol which were introduced by Fréchet in 1989 (*3*). These materials are prepared by the convergent growth approach in which the synthesis starts at the periphery of the macromolecule and proceeds radially inwards in a step-wise manner (*4*). As a result, larger and larger dendritic wedges, or dendrons, are constructed which are characterized by a very well defined monodisperse structure. A representative example, the fourth generation alcohol, [G-5]-OH, **1**, is shown in figure 1, and a number of important structural features are readily apparent. Firstly, the structure has a large number of surface functional groups (32 benzyl ethers), 31 internal 3,5-dioxybenzyl repeat units arranged in five layers, or generations, and a single hydroxymethyl group at the focal point of the dendron. In comparing the physical properties of dendrimers with linear polymers, the most relevant and realistic data will be obtained when the structural features of both molecules are exactly the same. For example, contemplation of the structure for the true linear analog of the fifth generation dendrimer, **1**, it should have 32 benzyl ether groups, a degree of polymerization of 31 and a single hydroxymethyl chain end. A two dimensional representation of such a linear polymer, **2**, is shown in figure 1 and comparison with the dendrimer clearly demonstrates that they are similar materials which differ only in the degree of branching. In fact the molecular formulae, $C_{441}H_{380}O_{63}$, and molecular weight (6680) of both molecules are exactly the same and so they are actual macromolecular isomers of each other.

In designing the synthesis of the linear analogs it became immediately apparent that a traditional exponential growth strategy(*5*) involving the synthesis of dimers, tetramers, octamers, etc. would not be applicable since the number of internal repeat units of the convergently grown dendrimers increases in a different numerical sequence. For example, both the exact linear analog, **2**, and the actual dendrimer, **1**, have 31 ($2^5 - 1$) internal repeat units which is not consistent with an exponential (2^n) approach. Therefore a strategy for the accelerated synthesis of monodisperse linear oligomers and polymers with 3, 7, 15, 31, 63, ($2^n - 1$) repeat units was developed (*6*). The basic strategy involves two parallel exponential growth strategies in which two series of oligomers, one with 2^n repeat units and the other with $2^n - 1$ repeat units, are prepared and then coupled together to give the desired linear analogs. A flowchart of the synthetic approach is shown in Scheme 1, the protected tetramer, **3 P-4-OH** (where P represents a phenacyl protected end group, 4 represents the number of repeat units, and OH represents a hydroxy chain end), of the 2^n series is deprotected with zinc in acetic acid to give the monophenol, **4 HO-4-OH** (where HO represents a phenol end group), which can then be coupled with the trimer, **5 L-3-Br** (where L designates this as an exact linear analog), to give the exact linear analog of the third generation dendrimer, **6 L-7-OH**, and subsequently other members of the $2^n - 1$ series. In parallel, the protected tetramer, **3 P-4-OH**, can also be brominated with CBr_4/PPh_3 to give the bromomethyl derivative, **7 P-4-Br**, and coupling with the monophenol, **4 HO-4-OH**, leads to the next generation octamer, **8 P-8-OH**. Repetition of these exponential growth strategies gives larger and larger linear dendritic analogs and was

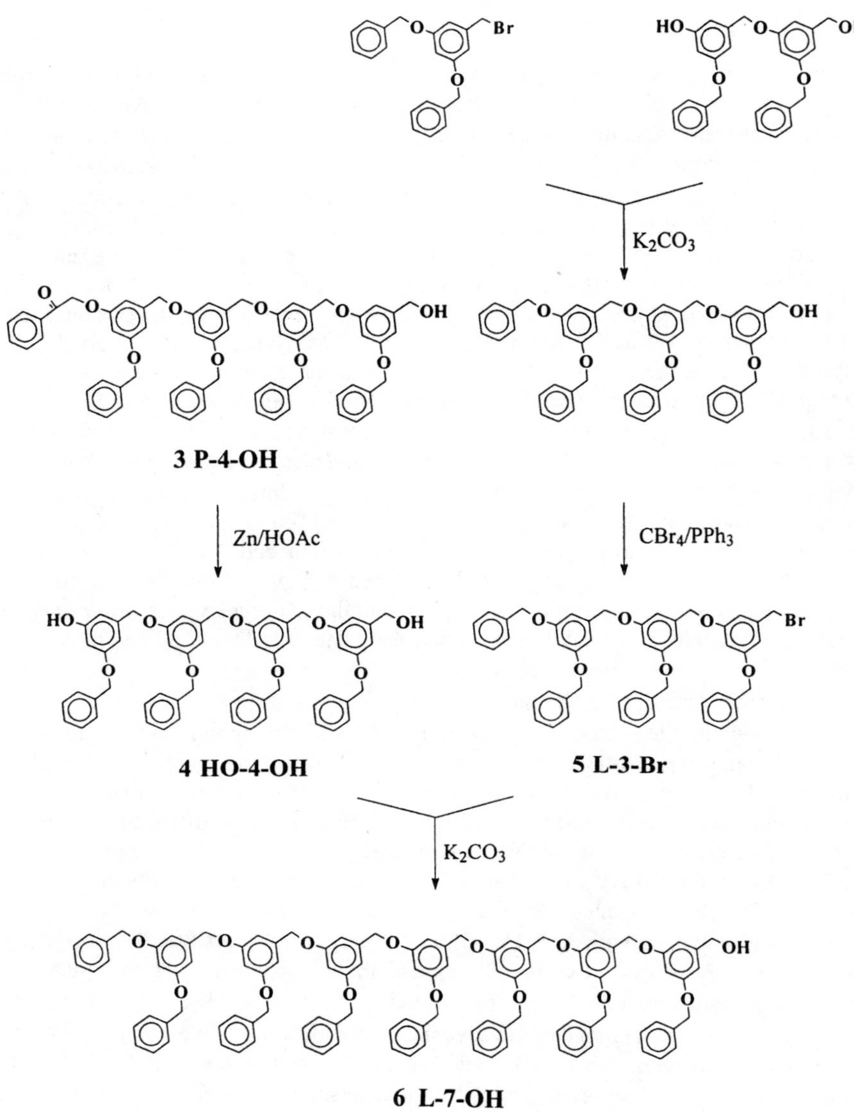

3 P-4-OH

4 HO-4-OH

5 L-3-Br

6 L-7-OH

Scheme 1

carried out till the sixth generation derivative, **9 L-[63]-OH**, which has 63 repeat units and a molecular weight of 13480.

Characterization of **9** and other linear analogs was performed using a variety of spectroscopic and chromatographic techniques. As expected the materials were essentially monodisperse and showed only a single molecular ion by MALDI-TOF at the desired molecular weight. However the ^1H and ^{13}C NMR spectra for the linear derivatives were considerably more complex than the corresponding dendrimer. For example, the tetramer, **4 HO-4-OH,** shows distinct resonances for each of the 8 benzylic groups, fully consistent with the total lack of structural symmetry (Figure 2). When the actual structures are considered this increased complexity for the linear analogs is expected since the random coil chains have lost all of the symmetry imposed by the perfectly branched architecture of the dendrimer. In fact, while it is difficult to successfully model these extremely large molecules and to represent them correctly in two dimensions, a comparison of minimized space filling models for the sixth generation dendritic and linear analogs clearly shows that the dendrimer is a much more compact and symmetrical structure than the random coil linear polymer.

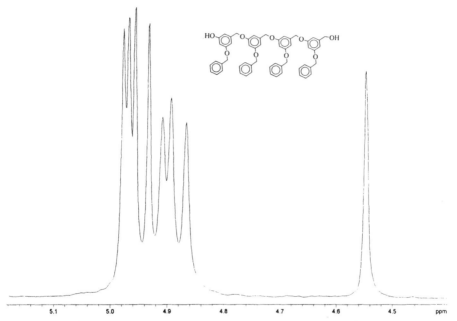

Figure 2. ¹H NMR of the tetramer, **4 HO-4-OH**, in CDCl₃ showing the region, 4.40-5.20 ppm.

Comparison of Physical Properties. The ultimate objective of these studies was to determine if there is any difference in physical properties between these two classes of macromolecular isomers. Significantly, a major difference was immediately observed while performing routine GPC analysis of these materials. As shown in figure 3, comparison of the retention times for the linear analogs versus the corresponding dendrimers displayed a remarkable trend. At low generation number, the retention time for the third generation dendrimer, **10,** is nearly the same as the linear analog, **11,** which suggests that in THF there is only a minor difference in hydrodynamic size. However on increasing the generation number a major difference in now observed with the retention time or hydrodynamic size (THF) of the linear analog, **12,** being significantly different, an approximately 40% larger volume when compared to the dendrimer, **13**. Similar results were observed in a variety of other solvents. These results are interesting from a variety of viewpoints, firstly it clearly shows that the highly branched architecture of the dendrimer leads to a much more compact and globular structure than the random coil architecture of its linear isomer. Secondly, and of particular importance, is the observation that differences in hydrodynamic volume only become readily apparent at generation 4 and start to be significant at generation 5 and above. Interestingly this discontinuity in behavior at around generation 4, coincides in general with changes in other dendrimeric properties(*7*) and may suggest that major differences in physical properties between linear and dendritic macromolecules only manifest themselves at higher generations.

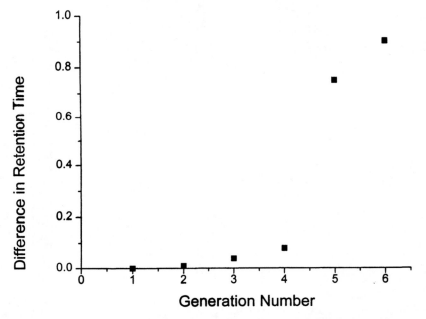

Figure 3. Plot of the retention time difference in minutes (linear – dendritic) versus generation number.

An excellent example of this discontinuity in properties is the relationship between generation number and the λ_{max} of a solvatochromic N,N-dialkyl-*p*-nitroaniline probe placed at the focal point, or core, of the dendrimers (*8*). As shown in Figure 4, UV-Vis measurements in a variety of low polarity solvents, in this case CCl_4, show that the λ_{max} of the solvatochromic probe increases with generation number, although in a non-linear fashion. A marked discontinuity is observed between generation 3 and 4 that has been correlated with a proposed shape transition from an extended to a more globular shape as the dendrimer grows larger. Therefore at higher generations the solvatochromic probe is shielded to a greater extent by the dendrimers and is influenced less by the low polarity solvent. If this rationale is true, then attaching a single solvatochromic probe to either the chain end or backbone of a linear

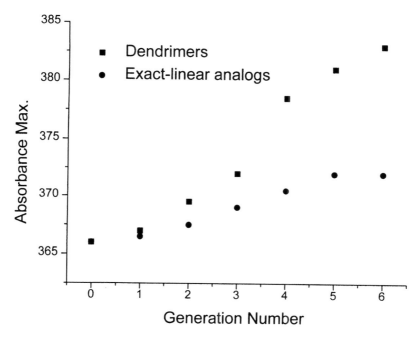

Figure 4. Plot of absorbance maximum versus generation number for solvaotochromically labeled dendrimers and linear analogs.

analog should lead to dramatically different behavior since it is unlikely that the linear polymer will be able to assume the same globular shape (Scheme 2). Interestingly, the behavior for the chain end functionalized linear analogs is indeed different; at low generation number the λ_{max} is similar to that observed for the dendrimer, however there is no marked discontinuity between generation 3 and 4, and the λ_{max} is observed to only increase slightly on going from G = 3 to G = 5 and actually plateaus at generation 5 and above. As a result, a significant difference is observed in the λ_{max} between the dendritic and linear analogs for generations 4-6, which is fully consistent with the diverging size and shape for these macromolecular isomers as they become larger and larger. While this change in behavior at ca. Generation 4 may be specific for benzylether dendrimers, it should be noted than a number of other groups have also observed changes in properties with generation number for other dendrimer families.

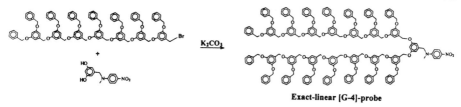

Scheme 2

For example, Moore has observed anomalous shifts in the fluorescence spectra of poly(arylacetylene) dendrimers with generation number and in this case a dramatic transition was observed between generation 4 and 5 in pentane (9).

This radically different size and shape for larger dendrimers can also have important effects on a number of other important physical properties. The influence of branching patterns on crystallinity and thermal properties was examined in detail for complimentary series of both the dendrimers and linear analogs. In conducting these experiments it was immediately obvious that the linear analogs were significantly more crystalline than the corresponding dendrimers. As reported previously (10), the fifth generation dendrimer, 1, is completely amorphous and shows a glass transition temperature of 315 K. This is in direct contrast to the linear analog, 2, which was found to be highly crystalline and displays a strongly exothermic melting transition at 423 K. This high level of crystallinity was also observed for all other linear polymers and is markedly different from their dendritic isomers, which are amorphous above generation 2. It should be noted that the linear analogs, when frozen into their glass have T_g's between 310-315 K, which is essentially the same as the values, found for the dendrimers. Similarly, both macromolecular architectures were shown to have the same thermogravimetric behavior, which is not unexpected since they are both composed of the same building blocks. These results suggest that basic thermal properties, such as T_g and T_{dec}, do not depend on the branching structure or macromolecular architecture of the sample and only depend on the number and nature of the repeat units. The level of crystallinity is however strongly dependent on the architecture of the polymer with the linear analogs being significantly more crystalline than the corresponding dendrimers. Again this is not unreasonable since it would be expected that the random coil linear chains could pack into a crystalline lattice much more easily than highly branched dendrimers. It should be noted, however, that in higher generations, dendrimers may become ordered again due to their increasingly sphere-like behavior. In this scenario, crystallization may be similar to the arrangement of colloidal particles in colloidal crystals. While this behavior has not been observed it is intriguing and may be a special feature of dendrimers.

The influence of branching was also evident when the solubility of dendrimers and their exact linear analogs was examined by vapor-liquid equilibrium experiments (11). As can be seen in Figure 5, the activity or solubility at 70°C in toluene of a series of dendrimers is markedly different from the corresponding linear analogs. All of the dendrimeric samples displayed increased solubility when compared to their linear analogs and this general relationship was also observed in a number of other solvents. While this difference in solubility may be due to different levels of crystallinity in the dendritic and linear isomers, it is also believed that the highly branched nature of the dendritic architecture plays a powerful role in determining the solubility of these materials. For example, the dendrimers are extremely soluble in a wide range of solvents such as THF, toluene, chloroform, etc., whereas the linear analogs are only sparingly soluble and can be recrystallized from these solvents. This increased solubility of dendrimers not only greatly facilitates their synthesis and processability but it also may play a major role in the future applications of these novel materials.

Similar behavior has been observed by Wooley in their comparison of the same dendrimers with linear analogs prepared by the one-step polymerization of 3-hydroxy-

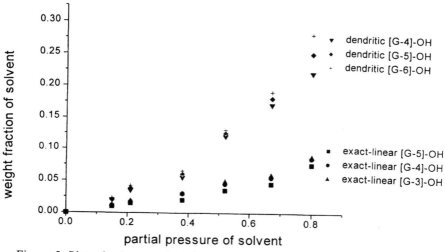

Figure 5. Plot of solvent weight fraction versus partial pressure of solvent for toluene at 70°C.

5-benzyloxybenzyl bromide followed by prep-GPC to obtain narrow polydispersity fractions (*12*). Interestingly, a direct comparison of these materials under a variety of micro-mechanical tests showed dramatically different behavior with the dendrimer machining in a fashion fully consistent with the lack of significant entanglements.

Hyperbranched Macromolecules. Because of their close structural and synthetic similarities to dendrimers, the comparison of hyperbranched macromolecules with their dendritic analogs has been studied by a number of groups. Ihre and Hult(*13*) have recently reported the synthesis and characterization of the dendritic analogs of the commercially available hyperbranched polyesters, Boltron®(*14*), based on 2,2-bis(hydroxymethyl)propionic acid, while Moore(*15*) has reported a novel solid-supported synthesis of hyperbranched poly(phenylacetylenes) which are the hyperbranched analogs of Moore's well studied phenylacetylene dendrimers(*16*). An intriguing study was also reported by Feast into the synthesis and physical properties of hyperbranched analogues of Tomalia's poly(amidoamine) dendrimers (*17*).

The most detailed studies have involved the comparison of linear, hyperbranched and dendritic analogs of aromatic polyesters based on 3,5-dihydroxybenzoic acid. In this case, the dendrimers were prepared by a step-wise procedure involving a series of DCC/DPTS coupling steps followed by deprotection of the focal point trichloroethyl ester group with Zn/HOAc (*18*). In contrast, the linear and hyperbranched isomeric polyesters were prepared by one-step polycondensation reactions and gave polydisperse materials with little or no control over molecular weights (*19*). This last feature demonstrates one of the drawbacks of this study, comparison of monodisperse, controlled molecular weight dendrimers with polydisperse linear or hyperbranched samples of different molecular weights does not allow many interesting and subtle features to be observed. However a number of observations were possible and the results obtained further confirms the unique differences associated with the perfectly branched nature of dendrimers (*20*).

Comparison of the solubility for the three different macromolecular architectures revealed that the dendrimer was the most soluble, followed by the hyperbranched derivative, and in turn both of these materials were dramatically more soluble than the linear analog. While it is difficult to draw strong conclusions from these studies it is worth noting than this series of compounds do not suffer from problems of partial crystallinity and so it can be concluded that the increased solubility is due to the branched architecture of the dendritic and hyperbranched derivatives. In concert with these results, it was also found that the thermal properties of the three different types of macromolecular isomers were essentially the same, again indicating that architecture does not influence T_g or T_{dec}. However the most intriguing feature of this study, and one which may point to a unique feature and possible application for dendrimers, is the difference in chemical reactivity for the functional groups located at the chains ends. These experiments demonstrated the clear reactivity advantage of dendrimers; reaction at the chain ends of non 'dense packed' dendrimers appeared to be relativity unimpeded and similar to small molecule chemistry, while reaction at the chain ends of the hyperbranched macromolecules were reduced and to a certain extent completely retarded for the linear analog. Recent results for a different family of dendritic macromolecules involving different chemistry also showed a similar structure-reactivity relationship between the 3 families of macromolecules (21). From these experiments it can be concluded that while both dendrimers and hyperbranched macromolecules can adopt globular, 3-dimensional shapes and have some similar properties, there are significant differences between the two families of highly branched polymers. Therefore, one should not assume that the properties observed for perfectly branched dendrimers would necessarily apply to hyperbranched macromolecules.

Conclusion

While this is not an exhaustive comparison of the physical properties of dendrimers versus other macromolecular architectures, such as linear polymers, it does suggest a number of important points. First and foremost is that dendrimers are indeed unique materials and their properties can differ substantially from their exact linear analogs. However, in most cases these differences only become apparent when the dendrimers becomes large enough to assume a true 'dendritic' shape; studies with small dendrimers should therefore be approached with caution since the structure may simply not be large enough to observe any unique properties. It should also be appreciated that not all physical properties will be different; for example, thermal properties are dominated by the nature of the repeat unit and, will be the same irrespective of the macromolecular architecture. Many physical and chemical properties are however sensitive to the branching pattern and while it has been conclusively shown that properties such as hydrodynamic radii, local nanoenvironment, crystallinity, and solubility are significantly different for dendrimers when compared to linear polymers, it is anticipated that further studies will show that a range of other properties such as rheology, interfacial activity, catalysis, diffusion, and importantly, chemical reactivity will also be unique for dendrimers.

Acknowledgements We would like to thank the National Science Foundation Materials Research Science and Engineering Center Grant DMR-9808677 for the Center for Polymeric Interfaces and Macromolecular Assemblies and the IBM Corporation for financial support of this work.

References

1. a) Fréchet, J.M.J., *Science* **1994**, *263*, 1710; b) Fréchet, J.M.J.; Hawker, C.J., in *Comprehensive Polymer Chemistry, 2nd Supplement* (Eds.: Aggarwal, S.L.; Russo, S.), Pergamon, Oxford, **1996**, pp. 77-133, c) Tomalia, D. A., *Adv. Mater.* **1994**, *6*, 529; d) Voit, B.I., *Acta Polym.* **1995**, *46*, 87; e) Newkome, G. R.; Moorefield, C. N.; Vogtle, F., *Dendritic Macromolecules. Concepts, Synthesis, Perspectives*, VCH, Weinheim, **1996**; e) Peerlings, H. W. I.; Meijer, E. W., *Chem. Eur. J.*, **1997**, *3*, 1563.

2. Hawker, C.J.; Farrington, P.; Mackay, M.; Fréchet, J.M.J.; Wooley, K.L., *J. Am. Chem. Soc.*, **1995**, *117*, 6123.

3. a) Fréchet, J.M.J.; Jiang, Y.; Hawker, C.J. Philippides, A.E. Proc. IUPAC Int. Symp., Seoul, 1989, pp. 19; b) Hawker, C.J.; Fréchet, J.M.J., *J. Chem. Soc., Chem. Commun.* **1990**, 1010.

4. Hawker, C.J.; Fréchet, J.M.J., *J. Am. Chem. Soc.* **1990**, *112*, 7638.

5. a) Zhang, J.; Moore, J.S.; Xu, Z.; Aguirre, R.A. *J. Am. Chem. Soc.* **1992**, *114*, 2273; b) Maddux, T.; Li, W.; Yu, L., *J. Am. Chem. Soc.* **1997**, *119*, 844; c) Tour, J.M., *Chem. Rev.*, **1996**, *96*, 537.

6. Hawker, C.J.; Malmström, E.E.; Frank, C.W.; Kampf, J.P., *J. Am. Chem. Soc.*, **1997**, *119*, 9903.

7. a) Moreno-Bondi, M.C.; Orellana, G.; Turro, N.J.; Tomalia, D.A. *Macromolecules* **1990**, *23*, 912; b) Naylor, A.M.; Goddard, W.A. III; Kiefer, G.E.; Tomalia, D.A. *J. Am. Chem. Soc.* **1989**, *111*, 2339.

8. Hawker, C.J.; Wooley, K.L.; Fréchet, J.M.J., *J. Am. Chem. Soc.*, **1993**, *115*, 4375.

9. Devadoss, C.; Bharathi, P.; Moore, J.S., *Angew. Chem., Int. Ed. Engl.*, **1997**, *36*, 1633.

10. Wooley, K.L.; Hawker, C.J.; Pochan, J.M.; Fréchet, J.M.J., *Macromolecules*, **1993**, *26*, 1514.

11. Mio, C.; Kiritsov, S.; Thio, Y.; Brafman, R.; Prausnitz, J.; Hawker, C.J.; Malmström, E.E., *J. Chem. Eng. Data*, **1998**, *43*, 541.

12. Fields, H.R.; Kowalewski, T.; Schaefer, J.; Wooley, K.L., *ACS Polym. Prep.*, **1998**, *39(2)*, 1169.

13. Ihre, H.; Hult, A.; Soderlind, E., *J. Am. Chem. Soc.*, **1996**, *118*, 6388.

14. Malmström, E.; Johansson, M.; Hult, A., *Macromolecules*, **1995**, *28*, 1698.

15. Bharathi, P.; Moore, J.S., *J. Am. Chem. Soc.*, **1997**, *119*, 3391.

16. Xu, Z.; Moore, J.S., *Ang. Chem. Int. Ed. Engl.*, **1993**, *32*, 1354.

17. Hobson, L.J.; Kenwright, A.M.; Feast W.J., *J. Chem. Soc., Chem. Commun.*, **1998**, 1877.

18. Hawker, C.J.; Fréchet, J.M.J., *J. Am. Chem. Soc.*, **1992**, *114*, 8405.

19.   a) Hawker, C.J.; Lee, R.; Fréchet, J.M.J., *J. Am. Chem. Soc.,* **1991**, *113*, 4583;
       b) Turner, S.R.; Walter, F.; Voit, B.; Mourey, T.H., *Macromolecules*, **1994**, *27*,
       1611; c) Wooley, K.L.; Hawker, C.J.; Lee, R.; Fréchet, J.M.J., *Polym. J.,* **1994**,
       26, 187.
20.   Wooley, K.L.; Fréchet, J.M.J.; Hawker, C.J., *Polymer,* **1994**, *35*, 4489.
21.   Trollsas, M.; Hawker, C.J.; Remenar, J.F.; Hedrick, J.L.; Ihre, H.; Hult, A., *J.
       Polym. Sci., Polym. Chem.*, **1998**, *36*, 2793.

Chapter 9

New *N*-Vinylformamide Derivatives as Reactive Monomers and Polymers

R. K. Pinschmidt, Jr., and N. Chen[1]

Corporate Science and Technology Center, Air Products and Chemicals, Inc., 7201 Hamilton Boulevard, Allentown, PA 18195

The utility of *N*-vinylformamide (NVF) as a functional monomer can be significantly expanded by base catalyzed addition of the amide nitrogen to Michael acceptors to generate novel derivatives with differentiated properties. Scope, synthesis conditions, reactions, and structure/property relationships for a series of alkyl 3-(*N*-vinylformamido)propionate Michael adducts and related 3-(*N*-vinylformamido)propionamides prepared on subsequent transamidation with alkanolamines and diamines are presented in the context of radiation cure coatings. Mechanistic details of the catalysis of the Michael addition reaction are discussed and rationalized.

N-Vinylformamide (NVF), a reactive functional monomer with novel physical and chemical properties and favorable toxicology (1,2) has shown significant promise in a number of application areas. It is highly reactive under free radical or cationic reaction conditions. The free radically prepared homopolymer is readily water soluble and can be hydrolyzed in a controllable fashion to give cationic or free base amine functional polymers. NVF, like other vinylamides, also copolymerizes well with most commercially available monomers, especially vinyl acetate, acrylamides, and acrylates (2). NVF's moderate toxicity enables the material to be used in applications where there is some possibility of worker exposure. However, it is difficult to obtain with NVF alone the broad range of physical and chemical properties needed for many applications requiring different glass transition temperatures and variable hydrophilic and hydrophobic characteristics. Also, there are situations where the toxicity profile of the monomer system is a primary concern and NVF may not be considered adequately safe.

[1]Current address: Henkel Corporation, 300 Brookside Avenue, Ambler, PA 19002–3498.

Over the past few years, we have developed a family of NVF derivatives prepared by base catalyzed addition of the acidic amide nitrogen to acrylates and related Michael acceptors (2-3). By employing the wide range of available Michael acceptors, a broad family of materials was prepared which show some very attractive properties. The derivatives tested to date have very low toxicity. Most are not water soluble and have much lower glass transition temperatures than NVF polymers. NVF and its derivatives also show significant similarities.   They are all radically co-polymerizable with major vinyl monomers, particularly acrylamides, vinyl esters, acrylates and maleates.   In particular, they are conveniently photocatalytically copolymerized with widely available acrylate based radcure components. Like NVF polymers, they can be hydrolyzed, but at much slower rates and to lower extents of conversion.

In this article, we will detail the properties of this group of materials, hopefully give the reader a tangible idea of the interesting properties and performance characteristics of the materials, and stimulate some thought on how to best use these materials in commercial applications.

Experimental

Synthesis procedures for representative NVF/(meth)acrylate Michael adducts and subsequent reaction with alkanolamines and diamines have been published (2). Standard methods for evaluation of products in radiation cure coatings were also described.

Di-[3-(N-vinylformamido)propionamido]methylcyclohexane and 1-[3-(N-vinyl-formamido)propionamido]-2-amino-methylcyclohexane. A 50 mL single-neck round bottom flask equipped with a distillation head was charged with 15.7 g (~0.1 mol) of methyl 3-(N-vinylformamido)propionate, 12.8 g (0.1 mol) of 1,2-diamino-3-methylcyclohexane and 0.10 g of 25% sodium methoxide in methanol solution. The mixture was stirred at 90 °C for 2 hr and 110 °C for an additional hour. Generated methanol was removed by distillation at reduced pressure to give a viscous liquid. NMR analysis indicated that about 58.5% of methyl 3-(N-vinylformamido)propionate was consumed.

Hydrolysis of MANVF homopolymer was very slow. At 65 °C with one equivalent of methanesulfonic acid in acetonitrile/methanol, 80% hydrolysis was achieved after ~75 hr (by titration). Base hydrolysis was 12% after 24 hr at reflux in MeOH and was accompanied by strong color build.

Copolymerization of methyl 3-(N-vinylformamido) propionate with NVF. A three neck round bottom flask, equipped with a mechanical stirrer, a water condenser, nitrogen inlet/outlet tube and an addition funnel, was charged with 15.4 g of MANVF, 0.2 g of Vazo 88, and 36.1 g of Dowanol (methoxypropanol). The mixture was heated to 100 °C, and stirred at that temperature for 30 min. A mixture of 4.6 g of NVF and 0.2 g of Vazo 88 in 10.5 g of Dowanol was added through the addition funnel in about 40 min. The mixture was heated continuously at 100 °C for four hr. A sample withdrawn from the reaction solution and analyzed by GC showed 76% of the MANVF and 98 % of the NVF were converted to polymer.

Radiation Cure Testing. Equal weight fractions of vinylamide reactive diluents and/or difunctional *N*-vinylformamido compounds were compared in model formulations (Table I).

Table I. Radcure Formulations

Component	Weight %
Epoxy diacrylate oligomer[a] or NVF based oligomer	50
TMPTA[b]	10
NVF diacrylate ester or diacrylate ester	10
N-vinyl reactive diluent[c]	30
Irgacure 184[d]	2.5 phr

[a]Ebecryl 3700 (UCB Radcure)
[b]Trimethylolpropane triacrylate (UCB Radcure)
[c]NVP, NVF or its Michael adducts
[d]1-Hydroxycyclohexyl phenyl ketone (Ciba-Geigy)

Films were drawn down on cleaned 5"x20" aluminum panels using a 10 mil drawing bar and cured under ultraviolet light in air using a commercial 300 watt/inch medium pressure mercury lamp and conveyor system (Fusion Systems) at 105 fpm and evaluated as described previously (2).

Results and Discussion

Monomer Synthesis. The synthesis of NVF derivative monomers is straightforward. Michael addition of NVF to an alkyl or substituted alkyl (meth)acrylate or other Michael acceptor is the core of the process (Scheme 1). The reaction can be catalyzed using common strong bases such as sodium or lithium methoxide. The reaction initiates at ambient temperature and is strongly exothermic.

Scheme 1. Synthesis of NVF Michael Adducts

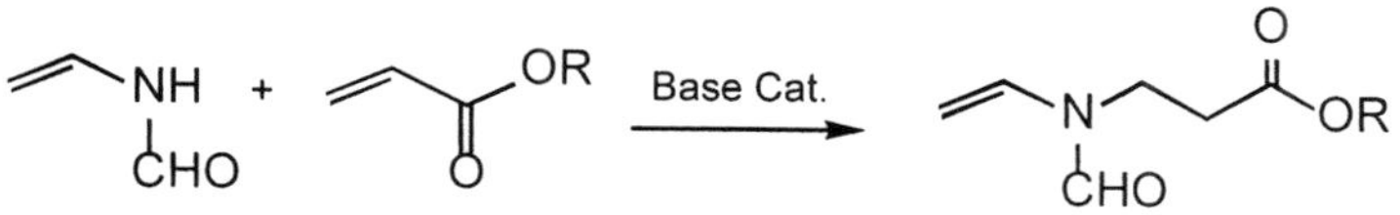

Usually, the reactions are complete within hours and give good conversions (40% - 90+%). In general, the larger the alkyl or other substituted alkyl group on the ester, the slower the reaction rate. Heteroatom containing substituents do not have a significant influence on the reaction rates and yields unless they contain active hydrogen (see below). A list of the derivatives prepared is shown in Table II.

Table II. Derivatives made of NVF and …

Allyl acrylate	tert-Butyl acrylate	Methyl acrylate
Benzyl acrylate	N, N-Dimethyl acrylamide	Lauryl acrylate
iso-Bornyl acrylate	N, N-Dimethylethyl acrylate	2,2,3,4,4,4- Hexafluorobutyl acrylate
Butyl acrylate	Ethyl acrylate	Methyl vinyl ketone
iso-Butyl acrylate	2-Ethyl hexyl acrylate	Iso-Octyl acrylate

Most derivatives can be purified efficiently by distillation under reduced pressure. The boiling points of some of the representative derivatives are depicted in Table III.

Table III. Boiling Points of Some NVF Michael Adducts

Methyl 3-N-vinylformamidopropionate	75 °C/0.8 torr
Ethyl 3-N-vinylformamidopropionate	88 °C/1 torr
Allyl 3-N-vinylformamidopropionate	105 °C/2 torr
Butyl 3-N-vinylformamidopropionate	96 °C/0.5 torr
t-Butyl 3-N-vinylformamidopropionate	80 °C/0.5 torr
i-Bornyl 3-N-vinylformamidopropionate	144 °C/0.7 torr
2-Ethylhexyl 3-N-vinylformamido-propionate	122 °C/0.8 torr
Lauryl 3-N-vinylformamidopropionate	155 °C/0.8 torr
1-N-Vinylformamidobutan-3-one	70 °C/0.8 torr
3-N-Vinylformamidopropionitrile	99°C/1.5 torr
Methyl 2-methyl-3-N-vinylformamido-propionate	76 °C/0.5torr

Variable results were observed with other Michael acceptors. Addition to methyl methacrylate was successful using metal hydrides (sodium was faster than calcium) or butyl lithium, while addition to crotonate or maleate esters failed in our hands. NVF reacted violently with acrolein, resulting in polymers.

NVF failed to react with Michael acceptors containing active hydrogens, i.e., acrylic acid, acrylamide or hydroxylethyl acrylate. The cause of the poor reactivity toward these compounds has not been demonstrated, but a mechanistic rational can be suggested. The key step in the reaction is clearly addition of a nitrogen anion of NVF to the Michael acceptor. The product anion (Scheme 2) appears not to oligomerize by adding additional acrylate, but to deprotonate additional NVF.

It is reasonable that weakly acidic materials could interrupt this cycle. If the initial deprotonation of NVF with alkoxide is slower than the reverse reaction, then alcohols will be inhibitors. This is consistent with the frequent observation of an induction period for the synthesis, especially using alkoxide base catalysts in alcohol,. The inhibiting alcohol should, however, be consumed, if only slowly, via addition to the Michael acceptor to yield alkyl ether. When ROH is finally consumed, the resulting anion should be basic enough to deprotonate NVF and the rate of the catalytic cycle leading to NVF Michael adduct should increase.

Detailed reaction conditions, product characterization and spectroscopic data have been described in previous publications (2,4).

Scheme 2. Michael Adduct Anion Reactions

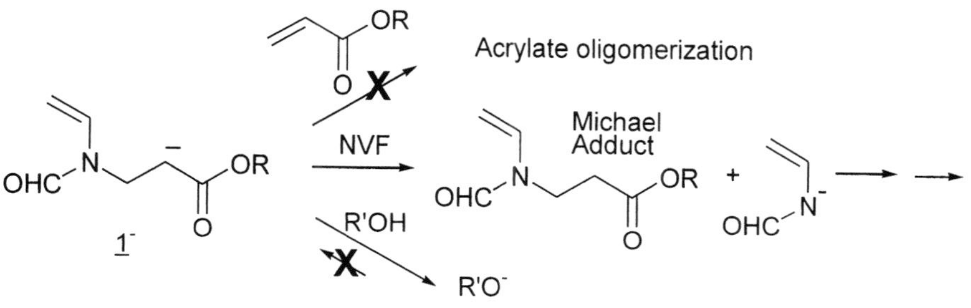

Scheme 3. Alcohol Depletion Mechanism

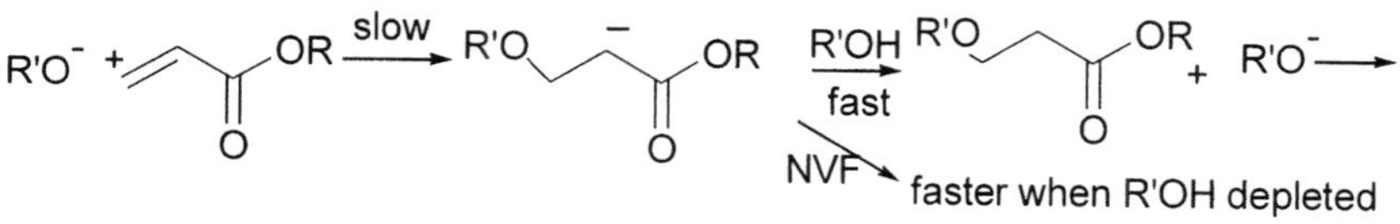

Hydroxyl Functional Monomers. Although it is difficult to prepare hydroxyl functional NVF derivatives using the NVF Michael addition reaction, the nonfunctional NVF/acrylate derivatives readily undergo transamidation reactions to yield a variety of new functional derivatives (Scheme 4).

Scheme 4. Transamidation of Michael Adducts

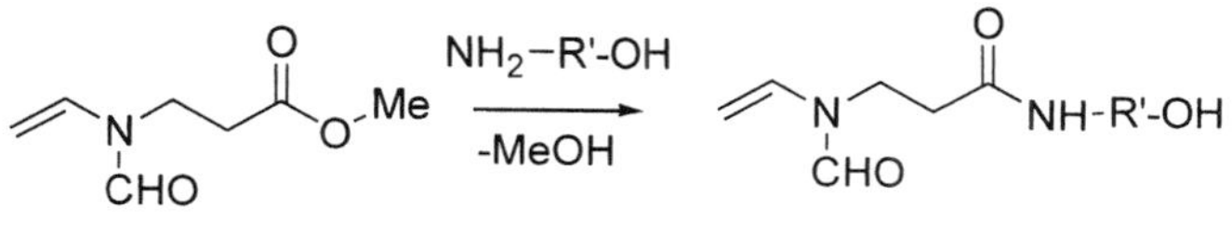

Successful examples include mono- and diethanolamine, diaminoethane, 1,2-diaminopropane, and hexamethylenediamine. Diaminoethane, even using 1:1 stoichiometry, gave exclusively the 2:1 adduct with MANVF. The product was highly crystalline; of the solvents tested, it had decent solubility only in alcohols. Curiously, when this crystalline adduct was melted and then mixed to give a typical radiation cure mixture (epoxy resin, crosslinker and reactive diluent), the mixture remained nicely liquid. 1,2-Diaminopropane successfully gave the 1:1 adduct in 90% yield after 5 hr at 115 °C (vs 90 °C/0.5 hr for other transamidations), but the higher temperature contributed color. Surprisingly, a number of other amines were either not successful (simple alkylamines such as mono- and diallylamine, aminobutyraldehyde dimethyl acetal), or required forcing conditions and gave poor yields and purities (N,N-dimethylethylenediamine, diethylenetriamine, 5-aminopentanol, and 1,2-diaminocyclohexane). The poor conversion of hindered primary amines in this reaction is easily rationalized as an issue of sterics, but the seeming requirement of active hydrogen elsewhere in the nucleophile to achieve useful conversions is surprising. One can draw transition states which include stabilization of developing charge during amine attack on the ester by a remote active hydrogen, but the picture is not compelling.

The successful preparation of hydroxyl functional NVF derivatives opened another option for preparing monofunctional resins for use as macromers or difunctional N-vinylformamide terminated resins for radiation cure applications (see below).

Difunctional NVF Derivatives

Diacrylate Michael Adducts. The synthesis of difunctional NVF Michael adducts from diacrylates works only under particular conditions. Most successful alkoxide catalysts for the monofunctional derivatives gave slow reaction rates and low conversions. Stronger bases such as butyl lithium did enhance the reaction rate, however, BuLi inevitably promoted anionic initiated polymerization as well and resulted in gel formation. It was found that sodium hydride gave the best overall result, giving good reaction rates and conversions to gel free difunctional products (Scheme 5).

Scheme 5. Synthesis of Difunctional Michael Adducts

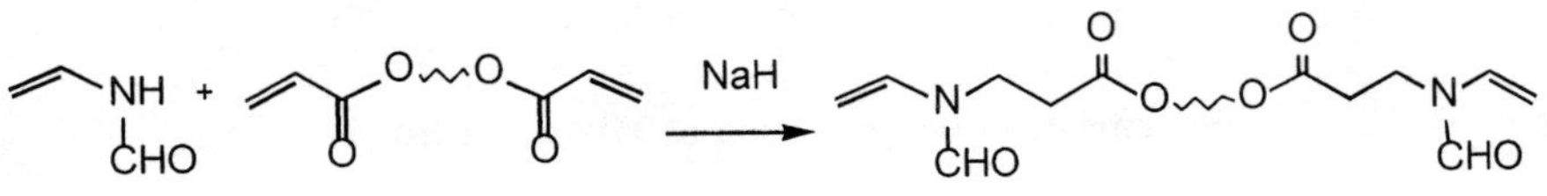

Two examples are the Michael addition of NVF with hexanediol diacrylate or butanediol diacrylate to give colorless moderate viscosity products (360 cp for NVF/hexanediol diacrylate). These products are very high boiling and distillative purification is at best a marginal option. Fortunately, good control of exotherm and precursor purity gives good color without distillation. Neutralization of catalyst residues and use of inhibitors is strongly advisable to suppress gellation on standing.

Di-N-vinylpropionamides. N-vinylformamide terminated reactive resins are readily prepared using the hydroxyl or amino functional NVF derivatives described in the previous section. Scheme 6 shows one option using diisocyanates as linkers. The resin can be readily cured and gave excellent properties as described in our earlier papers (2-4).

Di-*N*-vinylformylureas. A third option to prepare divinyl functional oligomers using NVF is through 1:1 reaction of NVF with a diisocyanate, such as isophoronediisocyanate, followed by reaction of the second isocyanate with a diol. As discussed earlier (2), the low reactivity and poor stability of the isocyanate adducts makes this option impractical.

Scheme 6. Synthesis of Divinyl Resins Using Diisocyanate Linkers

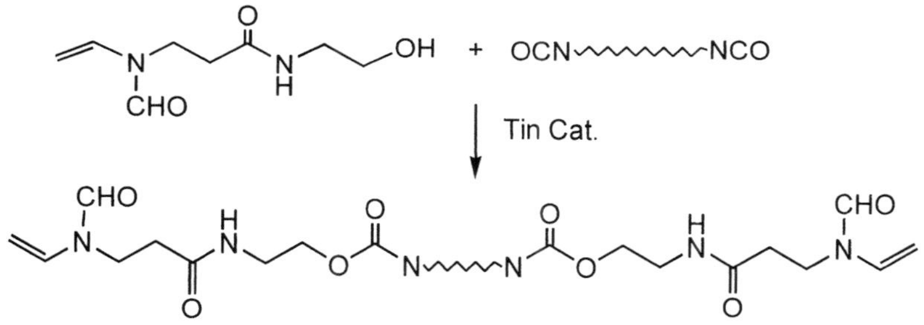

Properties of NVF Derivatives

1. Toxicity. One of the most striking features of this family of derivatives is their low toxicity. Table IV compares a typical NVF Michael adduct with its acrylate precursor.

Table IV. Toxicity Comparison

	NVF/Methyl acrylate Michael adduct	Methyl acrylate
Dermal LD$_{50}$	>2000 mg/kg	1243 mg/kg
Oral LD$_{50}$	>5000 mg/kg	277 mg/kg
Ames Assay	Negative	Negative
Sensitization	Negative	Positive
Eye Irritation	Positive	Positive

This unexpected feature is particularly desirable in applications involving the possibility of direct contact or exposure, such as radiation cure coatings.

2. Thermal and hydrolytic stability. NVF derivatives are notably more stable than NVF itself. Upon heating NVF/methyl acrylate Michael adduct at 60 °C for 96 hours, no significant viscosity increase was observed. NVF Michael adducts can be hydrolyzed under acidic conditions, but much more slowly than NVF. For example, when NVF/methyl acrylate Michael adduct was mixed with water in the presence of methanesulfonic acid (pH<1) and heated at 80 °C, it was hydrolyzed after 8 hours. The hydrolysis was slower when acetic acid was employed as the catalyst. Triethylamine or sodium bicarbonate (pH~9) give very slow hydrolysis: about 10% after 16 hours at 80 °C.

3. Water solubility. As expected, the water solubility of NVF derivatives decreased as the size of the alkyl group increased (Table V).

Table V. Water Solubility

NVF Derivative	NVF	MANVF	EANVF	BANVF	EHANVF
% H$_2$O Solubility	Soluble	24%	7%	~1%	<0.5%

4. Radical polymerization and co-polymerization. NVF derivatives are readily polymerized and co-polymerized through conventional free radically initiated reactions. For example, the polymerizations of NVF/methyl acrylate, NVF/ethyl acrylate, NVF/butyl acrylate, and NVF/ethylhexyl acrylate were carried out in Dowanol (methoxypropanol) at reflux (~100 °C) to give polymers with almost 90% conversion for all of them after five hours (Figure 1).

Polymers containing NVF derivatives have dramatically lower glass transition temperature than those made with NVF. Adducts of lower acrylates give the highest Tg while higher acrylates provide lower Tg (see Table VI).

Table VI. Michael Adduct Homopolymer Glass Transition Temperatures

Homopolymer	NVF	MANVF	EANVF
Tg (°C)	150	48	24

Copolymerizations with less hindered monomers such as acrylates, NVF, or vinyl acetate were faster and went easily to high conversion. Copolymerizations in methanol to partial conversions indicate reactivity ratios similar to those of NVF, with slightly faster conversion of either acrylates or NVF and slightly slower conversions of vinyl esters than the Michael adduct comonomers (Figure 2).

5. Cationic Oligomerization. Although no detailed study of cationic polymerization has been done for these NVF derivatives, when a strong acid such as methanesulfonic acid was added to these derivatives, significant viscosity increase was observed. This is an indication that a cationic catalyzed oligomerization process is involved similar to that observed for NVF (which, however, can react violently with strong acids). At ambient temperature in the presence of small amounts of weak

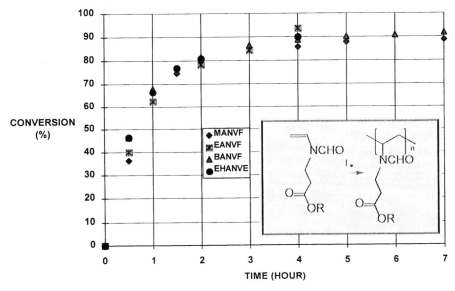

Figure 1. Conversion/Time Plot for Michael Adduct Homopolymerizations

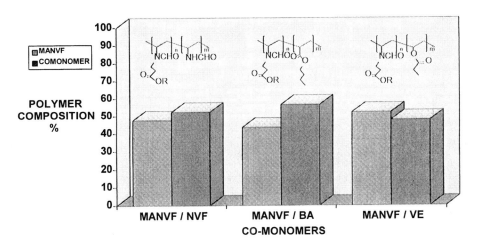

Figure 2. Copolymer Composition for Methyl *N*-Vinylformamidopropionate Copolymers

acid, such as 5% acrylic acid,, NVF/methyl Michael adduct remained stable with virtually no color change or viscosity increase for several days. However, the viscosity increased gradually when the acrylic acid concentration increased to >10 wt.% Usually, dark color developed as well.

6. Hydrolysis of NVF Michael Adduct Polymers. The polymers and copolymers prepared from these NVF derivatives can be hydrolyzed or alcoholyzed under proper conditions. However, at least the solvolysis is much more difficult than with NVF polymers At about 80 °C in the presence of 1.8 M methanesulfonic acid in a mixture of ethanol and acetonitrile the cleavage of the formamide followed second order reaction kinetics and had a half life of 12.5 hours.

7. NVF Derivatives as Reactive Diluents. The low viscosities, low vapor pressures, and good solvency of these NVF derivatives, combined with low toxicity, good storage stability and color, and ready variation of the hydrophobicity and Tg strongly suggest potential as reactive diluents in radiation cure formulations. Several preliminary evaluations illustrating this were presented previously (2). Adducts of lower acrylates give higher hardness and the lowest viscosities (and greatest viscosity reduction when used as reactive diluents), while higher acrylates provide softer films and higher viscosity. The heavier derivatives tended to show shallower depth of cure (DOC) as well (Table VII).

Scheme 7. Hydrolysis of Michael Adduct Polymer

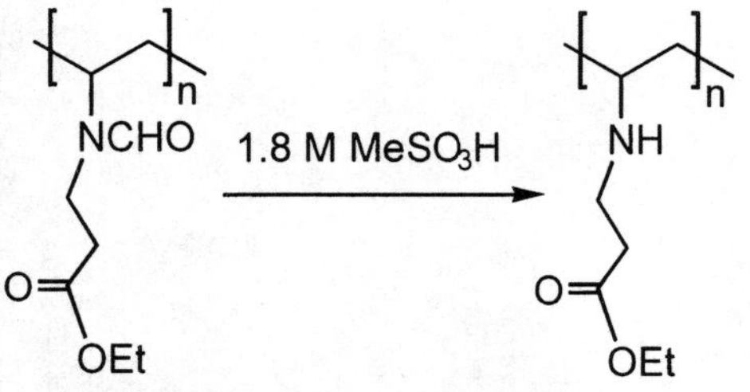

Table VII. Viscosity, Hardness and Depth of Cure

Vinylamide vs Property & performance	NVP	NVF	EA/ NVF	BA/ NVF	TBA/ NVF	EHA/ NVF
Formulation viscosity (cp, 25 °C)	464	625	1080	1250	2620	2220
Perzos hardness	252	255	233	160	208	166
Depth of cure (mil)	90	90	70	55	60	73

Performance was also a strong function of reactive diluent amount. Use of equimolar levels instead of equal weights gave lower viscosities, but also softer films and lower depth of cure. BANVF cured anomalously slowly in our urethane acrylate tests, but performed well in epoxy acrylate testing.

8. NVF Derivatives as Crosslinkers. Tables VIII and IX show that the experimental di-N-vinylformamido crosslinkers described above performed as well as a commercial crosslinker, except for a somewhat higher viscosity. Sixty degree gloss was >99 for the samples in Table IX.

Table VIII. Comparison of di-NVF and Diacrylate Crosslinkers Using NVF as Reactive Diluent

	TRPGDA	BDNVFP	HDNVFP
Formulation viscosity (cp)	162	272	220
Film thickness (mil)	4.0	4.0	3.4
Persoz hardness	255	191	258
Water double rubs	>200	>200	>200
MEK double rubs	>200	>200	>200

TRPGDA: tripropylene glycol diacrylate,
BDNVFP: butanediol diacrylate/ NVF di-Michael adduct,
HDNVFP: hexanediol diacrylate/NVF di-Michael adduct.

Table IX. Comparison of Di-NVF and Diacrylate Crosslinkers Using Methyl 3-(N-vinylformamido)propionate as Reactive Diluent

	TRPGDA	BDNVFP	HDNVFP
Formulation viscosity (cp)	330	708	564
Film thickness (mil)	4.0	3.6	3.6
Persoz hardness	159	146	152
Water double rubs	>200	>200	>200
MEK double rubs	>200	>200	>200

Table X shows the performance of the di-*N*-vinyl crosslinker made from diaminoethane and MANVF using the same formulation as above. Performance is again comparable, except for somewhat higher viscosity.

Table X. Comparison of Di-NVF Diamide and Diacrylate Crosslinkers

	TRPGDA	Di-NVF Diamide
Formulation viscosity (cp)	259	426
Film thickness (mils)	>4.0	> 4.0
Gloss (1 pass, 20 °/60 °/85 °)	103.9/107.9/99.8	106.4/109.7/99.8
Persoz hardness (1 pass)	286	269
Water double rubs	>200	>200
MEK double rubs	>200	>200

9. NVF Derivatives as Reactive Resins. Table XI compares the properties of formulations containing a di-*N*-vinylformamido-terminated urethane oligomer (NVFTO, described above) and a commercial urethane diacrylate oligomer (ATO, UCB's Ebercryl 3,700) with different monofunctional diluent monomers. As can be seen, the *N*-vinyl oligomer cured rapidly to give chemically-resistant crosslinked films that exhibited good gloss and performance equal to or superior to the commercial oligomer.

Table XI. Comparison of Di-NVF Terminated Urethane Oligomer and a Commercial Urethane Based Oligomer (ATO)

Oligomer	NVFTO		NVFTO		ATO
Diluent	NVF		MANVF		MANVF
Formulation visc (cps)	390	390	720	720	1700
Film thickness (mil)	3.1	4.0	4.0	3.8	4.0
# Passes @ 105 fpm	1	4	1	4	2
Persoz hardness	38	109	68	64	38
Gloss (60°)	99.7	98.2	99.0	99.6	99.7
MEK double rubs	>200	>200	>200	>200	>200

Conclusions

N-Vinylformamide, in addition to its utility as a reactive, water soluble precursor to cationic or reactive amine functional polymers, shows rich chemistry for the synthesis of new derivatives by reaction at the acidic nitrogen. This chemistry is the basis of new vinyl and divinyl monomers with widely and easily varied hydrophobicity and derived polymer Tg and attractively low toxicity. The derivatives exhibit facile free radical reactivity in homo- and copolymerizations. They also show good thermal and chemical stability and very low viscosity. Although the Michael addition approach cannot be used with Michael acceptors such as hydroxyethyl acrylate containing active hydrogen, they are reactive via ester transamidation with certain diamines and alkanolamines to give new hydroxy- or amine-functional vinyl monomers. These novel NVF derivatives offer promise in particular as reactive diluents, crosslinkers, and oligomer resin precursors for radiation cure coatings.

Acknowledgments

Of the many people who have worked in this area, particular mention must be made of W. E. Carroll, J. N. Drescher-Reidy, K. B. Kelby, W. L. Renz, J. van Horn, A. Kotz, F. Prozonic, and D. J. Nagy who contributed in significant measure to the laboratory, analytical support, and commercial work.

Literature Cited

1. K. Pinschmidt, Jr. and D. J. Sagl in Polymeric Materials Encyclopedia, J. C. Salamone, ed.-in-chief, CRC Press: New York, New York, 1996, pp 7095-7103.
2. Pinschmidt, R. K. Jr.; Renz, W. L.; Carroll, W. E.; Yacoub, K.; Drescher, J. N.; Nordquist, A. F., Chen, N. *J. Macromol. Sci. - Pure and Appl. Chem.* **1997,** *A34(10)*, pp 1885-1905.
3. Chen, N.; Renz, W. L.; Carroll, W. E.; Drescher, J. N.; Pinschmidt, R. K. Jr.; *4th North American Research Conference on Organic Coatings: Science and Technology; Nov. 1996.*
4. R. K. Pinschmidt, Jr. and N. Chen, *Polym. Preprints,* 39(1), **1998,** pp 639-640.

Chapter 10

Epoxy-Based Thermoplastics: New Polymers with Unusual Property Profiles

Jerry E. White[1], David J. Brennan[2], H. Craig Silvis[2], and Michael N. Mang[2]

[1]Epoxy Products and Intermediates Research and Development, The Dow Chemical Company, Freeport, TX 77541
[2]Corporate Research and Development, The Dow Chemical Company, Midland, MI 48674

Stoichiometrically-balanced polymerizations of diglycidyl ethers with difunctional amines, bisphenols, difunctional sulfonamides, dicarboxylic acids or dithiols yield a new family of thermoplastic resins with a rich array of physical properties. Of particular interest, epoxy thermoplastics often have outstanding barrier to oxygen and other atmospheric gases; this feature, along with their mechanical integrity and fabrication characteristics, suggest the use of these materials in protective packaging.

Over the last several decades, a major materials industry centered on epoxy thermosetting resins has been built on the lability of the oxirane or epoxy functionality to nucleophilic attack by amines, carboxylates and other species (*1*). Because of this fundamental characteristic, epoxy resins such as **1** can be cured with multifunctional nucleophiles in a variety of environments to yield crosslinked resins found in powder coatings, solvent-free and solvent-borne coatings, composites for electrical laminates and two-part adhesives. Epoxy oligomers also are used heavily in coatings for food and beverage cans, an application which reflects both the strong adhesion of these systems to metal and the ability of crosslinked epoxy resins to provide coated articles with outstanding corrosion resistance (*1*).

1   **2a**

Despite the spectacular success of epoxy-based materials in the thermoset arena, *thermoplastic* epoxy polymers have received comparatively little attention. Reactions of equimolar ratios of bisphenols with epichlorohydrin to afford poly(hydroxy ethers) such as **2a** were reported in 1963 (*2,3*), and polymer **2a** actually is an item of commerce sold primarily as a solvent-borne specialty coating resin (*4*). Even so, only recently has a focused study of stoichiometrically-balanced polymerizations of bisepoxides with dinucleophilic monomers shown that such reactions yield thermoplastic polymers (eq 1) with a broad and unusual array of properties (*5,6*). Of special interest, epoxy thermoplastics have generally excellent barrier to atmospheric gases (*2-6*), a partial consequence of the strong contribution of their pendent hydroxy groups to intermolecular cohesion via interchain hydrogen bonding (*7,8*). In addition, these new materials often exhibit a remarkable combination of clarity, mechanical integrity, and good processing characteristics, a property profile that makes the resins attractive for protective packaging as well as for other applications.

$$\text{(structure)} \quad + \quad \text{HAH} \quad \longrightarrow \quad \text{(polymer structure)}_n \qquad (1)$$

Synthesis of Epoxy Thermoplastics

Poly(hydroxy amino ethers). Primary amines or secondary diamines react cleanly with pure diglycidyl ethers such as **1** at 150 to 180 °C in solution or in the melt to yield soluble, thermoplastic poly(hydroxy amino ethers) (**3**) typified by the backbone sequences shown in Table I (*9-12*). These polymers are genuinely high-molecular-weight species; inherent viscosities (IV) of **3** in dimethylformamide (DMF) normally are well above 0.4 dL/g, and most have weight-average molecular weights (Mw) of 50-100,000 (SEC; relative to polystyrene standards). Also, light scattering/osmometry measurements indicate that polymer **3a**, derived from ethanolamine and diglycidyl ether **1** in solution, has an absolute Mw of about 60,000 with a polydispersity (Mw/Mn) of around 3.5 (*11*). Variations of Mw/Mn in **3** above the ideal value of 2 for a step-growth polymer (*13*) indicate that some branching accompanies linear chain growth in **3**, probably via reactions of a few zwitterionic alkoxides (**4**; necessary intermediates in the formation of **3**) with epoxy functionalities as shown in eq 2 (*9*). Such side reactions must be rare, however, since significant involvement of eq 2 during polymerization of amines and bisepoxides would disrupt the stoichiometry of the polymerization to afford low-molecular-weight or possibly crosslinked products rather than the soluble thermoplastics that are observed.

Initiated Polymerizations of Bisepoxides and Dinucleophiles.

In a group of especially versatile reactions, various arene diglycidyl ethers in ether solvents can po-

$$(2)$$

Table I. Synthesis of Poly(hydroxy amino ethers)

$$1 \quad + \quad \text{HAH} \quad \longrightarrow$$

3

No.	-A-	IV, dL/g[a]	Tg, °C[b]
3a	—N— / OH	0.57	81
3b	—N N—	0.85	100
3c	Me \ —N—CH₂CH₂—N— / Me	0.59	49
3d	—N— / OMe	0.46	86

[a] 0.5 g/dL; DMF; 25 °C. [b] Differential Scanning Calorimetry (DSC).

lymerize at 100 - 200 °C with a broad range of simple and amide-containing bisphenols (*14-20*), difunctional sulfonamides (*21, 22*), dicarboxylic acids (*23-25*), and arenedithiols (*26*) to afford the corresponding poly(hydroxy ethers) and poly(hydroxy amide ethers) (**2**), poly(hydroxy ether sulfonamides) (**5**), poly(hydroxy ester ethers) (**6**) and poly(hydroxy ether sulfides) (**7**) outlined in eq 3. While these transformations follow the stoichiometric dictates of a step-growth process (*i.e.* an equimolar ratio of reactants is necessary to achieve high molecular weight in **2, 5-7**), they also encompass the elements of a chain process (*21*) in which the polymerization is initiated by formation of a reactive species that is continually regenerated until the polymerization is complete (see eqs 4 and 5). Effective initiators for the hybrid step-growth reactions of eq 3 include phosphonium and ammonium acetates, halides and hydroxides.

No.	HAH	-A-
2	HO–Ar'–OH	—O-Ar'-O—
5	$\overset{R}{HN}-\overset{O}{\underset{O}{S}}-Ar'-\overset{O}{\underset{O}{S}}-\overset{R}{NH}$	$-\overset{R}{N}-\overset{O}{\underset{O}{S}}-Ar'-\overset{O}{\underset{O}{S}}-\overset{R}{N}-$
	or	or
	$\overset{NH_2}{\underset{Ar'}{O=S=O}}$	$\overset{-N-}{\underset{Ar'}{O=S=O}}$
6	$HO_2C-R-CO_2H$	$-O_2C-R-CO_2-$
7	HS–Ar'–SH	—S–Ar'–S—

Mechanistically, formation of poly(hydroxy ethers) (**2**) from bisphenols and diglycidyl ethers can be rationalized by the sequence depicted in eqs 4 and 5 (*16,21*). The anion, X⁻, of a quarternary phosphonium or ammonium salt initiates polymerization by attacking an epoxide group to afford alkoxide **8**, which subsequently deprotonates a phenolic functionality to produce a reactive, nucleophilic phenate (**9**) along with neutral acetoxy- or halohydrin **10** (eq 4) . Reaction of phenate **9** thus formed with an epoxy moiety yields alkoxide segment (**11**), which propagates the polymerization when neutralized by an unreacted phenol group to form hydroxy propylene backbone linkage **12** and to regenerate a phenate (**9**) capable of condensing with unconsumed epoxy end groups and continuing chain growth. The reaction ultimately is terminated upon complete consumption of phenol or epoxy functions, which, when provided in stoichiometric balance, yield high-molecular-weight **2**. A practical consequence of eqs 4 and 5 is that acetates must be used at very low levels (< 1 - 2 mol %). During the initiation step (eq 4), acetate anions attack epoxide groups to irreversibly produce acetoxyhydrin species (**10**; X = AcO-), which act as chain terminators; thus, relatively high concentrations of acetate initiators lead to low-molecular-weight products (*16*). Indeed, quarternary ammonium halides generally are preferred initiators not only for preparation of polymers **2** but also for the synthesis of poly(hydroxy ester ethers) (**6**) and poly(hydroxy ether sulfides) (**7**). Since halo moieties are relatively good leaving groups, halohydrin end caps (**10**, X = Cl- or Br-) are transient species and can continue to participate in the polymerization (and thus step-wise chain growth) by reversion to an epoxy functionality or through direct halide displacement by phenate, thiolate or carboxylate (*16, 23*). For polymerizations of difunctional sulfonamides with bisepoxides, tetraalkylammonium hydroxides typically yield poly(hydroxy ether sulfonamides) (**5**) with higher molecular weight than is the case with halide initiators; in this instance, the reaction probably is initiated by direct deprotonation of a sulfonamide moiety rather than by a pathway like that of eq 4 (*21*).

$$\text{(4)}$$

$$\text{(5)}$$

Preparation of poly(hydroxy ethers) (2) is usually accompanied by some branch formation that probably results from attack of alkoxide intermediates 11 on epoxy chain ends in a process analogous to that of eq 2. For example, light-scattering/SEC studies of polymer 2a, prepared with an absolute Mw of 130,000 from the reaction of bisphenol A with diglycidyl ether 1, indicate that the polymer has about four branches per weight-average molecule (16). Structurally similar branching is likely present in poly(hydroxy ether sulfonamides) 5 (21) and poly(hydroxy ester ethers) 6, as well. Also, formation of 6 is further complicated by intramolecular trans-esterfication of the ester-ether alkoxide segments (13; eq 6), the presumed intermediates in the polymerizations of diacids with diglycidyl ethers. As a result, about 20 mol % of the backbone ester-ether linkages contain the pendant methylol residue of segment 14 while the remainder have expected structure 15 (see eq 6; for convenience, 15 is the only linkage depicted in structures of 6 shown in this chapter) (23-25). As will be discussed elsewhere, polymers 6 also can undergo intermolecular transesterification during or after polymerization; nonetheless, 6 are generally melt-stable resins which, like other epoxy thermoplastics, can be conventionally fabricated by a number of methods (23).

Properties of Epoxy Thermoplastics

Polymer Morphology. Most epoxy thermoplastics are amorphous, glassy polymers with glass transition temperatures (Tg) that range from about 25 °C to over 200 °C (5,6). As is the case with all thermoplastics, Tg in polymers 2, 3 and 5-7 are heavily influenced by the torsional mobility of the polymer backbone. For instance, in a series of poly(hydroxy amide ethers) (**2b-e**; Table II), derived from various arylene diglycidyl ethers and N,N'-bis(3-hydroxyphenyl)adipamide (**16**), Tg clearly climbs from 92 °C for **2b** to 146 °C for particularly immobile polymer **2e** with increasing steric bulk of the arylene linkage incorporated into the product chain (*27*). Not surprisingly, Tg of other amorphous epoxy thermoplastics respond similarly to analogous variations in primary structure (*10,14-16,21,23*).

16

Table II. Representative Poly(hydroxy amide ethers)

No.	-Ar-	IV, dL/g[a]	Tg, °C[a]	O$_2$TR, BU[b]
2b		0.43	92	0.5
2c		0.62	103	0.8
2d		0.37	118	1.0
2e		0.47	146	1.2

[a]See footnotes to Table I. [b]Values, which represent oxygen transmission rates (O$_2$TR) as cc-mil/100 in^2-atm(O$_2$)-24 hr, were determined according to ASTM method D3985-81 operated at about 65 % relative humidity and at 23 °C.

As a class, epoxy thermoplastics generally resist crystallization, but incorporation of biphenylene, *p*-phenylene or α-methylstilbene segments into backbones that are sufficiently mobile to allow ordered chain packing (*i.e.* Tg < 100 °C) can in a few instances yield polymers which exhibit true, endothermic melting transitions (Tm) at temperatures as high as 246 °C (*5,6*). For example, upon annealing, biphenylene-containing poly(hydroxy amide ether) **2f**, isolated as a transparent, amorphous glass with a Tg of 110 °C, crystallizes to a tough, opaque plastic which melts near 190 °C and displays an anticipated elevation of its Tg to 117 °C (Table III) (*27*). Similarly, introduction of *p*-phenylene linkages into poly(hydroxy ester ethers) like **6d** (Tm = 72 °C) or poly(hydroxy amino ethers) such as **3e** (Tm = 204-246 °C) results in polymers with semi-crystalline morphologies (*9*).

Table III. Comparison of Amorphous and Crystalline Epoxy Thermoplastics

$$\left[\!\!-O\diagup\overset{OH}{\diagdown}A\diagup\overset{OH}{\diagdown}O\text{-Ar}\!\!-\right]_n$$

No.	-A-	-Ar-	Tm(Tg), °C[a]	O$_2$TR, BU[a]
2f	$-O$-(m-C$_6$H$_4$)-HN\overset{O}{C}-(CH$_2$)$_4$-\overset{O}{C}NH-(m-C$_6$H$_4$)-$O-$	biphenylene	-- (110)	0.30
2f[b]	$-O$-(m-C$_6$H$_4$)-HN\overset{O}{C}-(CH$_2$)$_4$-\overset{O}{C}NH-(m-C$_6$H$_4$)-$O-$	biphenylene	190 (117)	0.06
2a	$-O$-(C$_6$H$_4$)-C(Me)(Me)-(C$_6$H$_4$)-$O-$	Me,Me-isopropylidenediphenylene	-- (100)	9.0
2g	$-o$-(C$_6$H$_4$)-C(Me)=CH-(C$_6$H$_4$)-$o-$	Me-stilbene	190 (81)	0.40
6a	$-O_2C-(CH_2)_4-CO_2-$	Me,Me-isopropylidenediphenylene	-- (45)	3.0
6b	$-O_2C-(CH_2)_4-CO_2-$	Me-stilbene	149 (44)	0.5
6c	$-O_2C-(CH_2)_{10}-CO_2-$	Me-stilbene	139 (34)	0.60
6d	$-O_2C-(CH_2)_{10}-CO_2-$	p-phenylene	72 (10)	10.0

[a]See footnotes to Tables I and II. [b]Annealed.

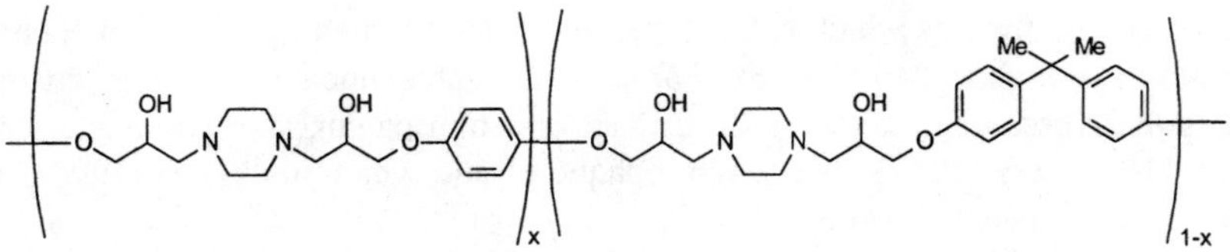

3e

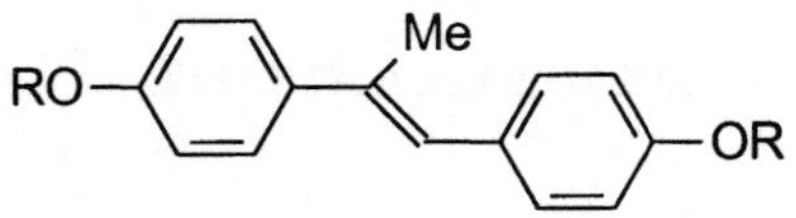

17a, R =

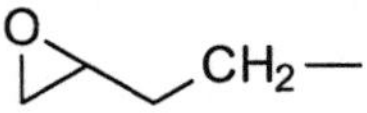

17b, R = H

In addition to classically semi-crystalline polymers like **2f**, **3e** and **6d**, liquid crystalline epoxy thermoplastics are accessible by polymerization of various dinucleophilic monomers with diglycidyloxy-α-methylstilbene (**17a**) and related materials (*28*). As an illustrative example, reaction of aliphatic diacids with **17a** afford liquid crystalline poly(hydroxy ester ethers) such as **6c** (Table III), a polymer that melts to a nematic state at 139 °C and ultimately becomes isotropic at temperatures above 200 °C. A possibly more intriguing material, the poly(hydroxy ether) (**2g**; Table III) formed from reaction of **17a** with the corresponding bisphenol (**17b**), apparently is a monotropic liquid crystalline thermoplastic (*29*) that enters an ordered melt only when super cooled below its Tm of 190 °C (*28*). Note that the structure of **2g** is less regular than Table III implies; isomerization of the stilbene double bond in **17b** during polymerization generates *cis* linkages (roughly 10 mol %) along the poly(hydroxy ether) chain. Also, since the α-methylstilbene species is unsymmetrical, the backbone of **2g** can be populated by any and all of the three diads, **2g'**, **2g"**, and **2g'''**. A similar argument can be applied to the backbone sequence in polymer **6c**, which also contains about 20 mol % of the methylol-containing linkage (**14**) that, as noted above in eq 6, is characteristic of all poly(hydroxy ester ethers).

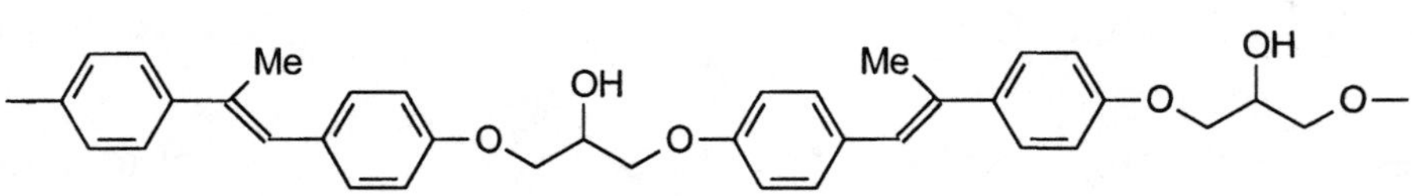

2g'

2g''

2g'''

Barrier Properties of Epoxy Thermoplastics. The permeability (P) of a polymer to an atmospheric gas is related to both the solubility (S) and diffusion rate (D) of the gas in and through the matrix according to the simple relationship, $P = DS$. Along with environmental factors like temperature and humidity, influences on D and S for thermoplastics include 1) the level of crystallinity in a resin; 2) any structural features of a polymer that contribute to interchain cohesion (cohesive energy density) via polar interactions; and 3) backbone geometries that affect intermolecular packing efficiency and free volume (*7,8*). In polymers with good barrier properties (*i.e.* very low permeability), crystallinity, interchain cohesion and effective packing work in concert to control the permeation rate of oxygen and other small gas molecules, and distinguishing the relative importance of these factors often can be difficult (*8*). However, the effect of intermolecular hydrogen bonding potential on interchain cohesion and ultimately on the barrier characteristics of epoxy thermoplastics can be at least partially isolated from other factors and in many cases is quite striking. For example, Table IV reveals that for relatively unoriented, compression-molded films of a group of poly(hydroxy amino ethers) (**3**) that are amorphous, and thus morphologically identical, the oxygen transmission rate (O$_2$TR) at room temperature drops dramatically from 6.3 cc-mil/100 in^2-atm (O$_2$)-24 hr (expressed here as Barrier Units or BU) to 0.8 BU when the apolar propyl residue of **3f** is replaced with the hydroxyethyl substituent of **3a**, which obviously can contribute to increased intermolecular hydrogen bonding. Substituting the bisphenol A nucleus in **3a** with a more compact *m*- or *p*-phenylene segment in poly(amino ethers) **3g** and **3h** further increases the number of hydrogen bonding sites per unit chain length in this series of polymers with a consequential additional decrease in O$_2$TR to 0.3 and 0.04 BU (Table IV); the later value is, we believe, the lowest O$_2$TR ever recorded for a completely amorphous, unoriented thermoplastic (*18*) and is comparable to that of the crystalline ethylene-vinyl alcohol copolymers used commercially in multilayer rigid and flexible packaging (*30*). With carbon dioxide transmission rates (CO$_2$TR) of 0.8 - 3.9 BU, polymers **3** also have remarkable barrier to CO$_2$; for comparison, bottle-grade poly(ethylene terephthalate) commonly used in plastic soft drink bottles has a CO$_2$TR of 25 - 30 BU (*30*). Because poly(hydroxy amino ethers) **3** combine such barrier performance with clarity and transparency, robust mechanical profiles and good fabrication characteristics (see below), these materials have real promise in protective packaging applications.

While molecular packing can also play a role in determining the O_2TR of epoxy thermoplastics, effects of pendent groups on the permeability of these polymers are less conspicuous than one might intuitively anticipate. For the series of amorphous poly(hydroxy amide ethers) (2b-e) shown in Table II, increasing the size of substituents on these otherwise identical polymers by replacing the arylene methylene linkage of 2b with an isopropylidene segment (2c), a phenylmethylmethylene unit (2d) and finally a bulky cardo fluorene species (2e) prompts measurable, but nevertheless, small increase in O_2TR from 0.5 to 1.2 BU. Other structurally analogous epoxy thermoplastics behave similarly (10,14-16,21,23); indeed, among glassy epoxy thermoplastics, only polymers with exceptionally disruptive side chains exhibit O_2TR above 10 BU (4-6).

Table IV. High-Barrier Amorphous Poly(hydroxy amino ethers)

No.	R	-Ar-	Tg, °C[a]	O_2TR, BU[a]	CO_2TR, BU[b]
3f	Me	(Me-C-Me, diphenyl)	53	6.3	---
3a	OH	(Me-C-Me, diphenyl)	81	0.8	3.9
3g[c]	OH	(p-phenylene)	52	0.3	0.8
3h	OH	(m-phenylene)	56	0.04	1.4

[a]See footnotes to Tables I and II. [b]Determined at 0 % relative humidity. [c]Unlike some p-phenylene containing polymers which readily crystallize (see text), polymer 3g is totally amorphous (9).

Because oxygen or other gaseous molecules are unlikely to dissolve in the crystalline portions of an ordered thermoplastic and diffusion is limited to the amorphous regions of the polymer matrix, such materials present a tortuous path to diffusing species which results in gas transmission rates that are lower than those observed in amorphous resins having similar polarity and packing efficiency (7,8). Thus, as Table III indicates, the annealed, crystalline version of poly(hydroxy amide ether) **2f** has an O_2TR (0.06 BU) that is notably lower than that of its amorphous counterpart (0.3 BU), a five-fold decrease in permeability that can only be ascribed to the introduction of crystallinity since the annealed material is unchanged in all other respects. Also, in the group of poly(hydroxy amino ethers) **3e** described above, O_2TR drops from 0.39 BU for a composition (x = 0.67) with low levels of crystallinity (reflected by a ΔHm of only 3 J/g) to 0.04 BU for highly crystalline **3e** (x = 1.0; ΔHm = 56 J/g), additional evidence for the positive impact that crystalline morphologies can have on the barrier performance of epoxy thermoplastics (9).

Other amorphous and semi-crystalline epoxy thermoplastics with similar primary structures exhibit readily observable differences in O_2TR (although in these cases, relative barrier performance may also reflect variations in packing efficiency in amorphous regions of the resins). For instance, unoriented films of α-methylstilbene-containing, liquid crystalline poly(hydroxy ester ethers) **6b,c** display O_2TR (0.5-0.6 BU) that are five-to-six times lower than that of a film of amorphous analog **6a** (O_2TR = 3.0 BU; Table III), the polymeric adduct of adipic acid and diglycidyl ether **1**. The difference in permeability of commercially available poly(hydroxy ether) **2a** and structurally similar, liquid crystalline polymer **2g** is even more dramatic; the O_2TR of **2g** (0.4 BU, unoriented) is over twenty times lower than that of amorphous **2a** (Table III), and the CO_2TR of **2g** also is an impressively low 1.5 BU (28). Note that permeation rates for oxygen and carbon dioxide in these liquid crystalline polymers will almost certainly decrease further in oriented specimens (7,8), but stretched or blow molded samples of **2g** and **6b,c** have not yet been prepared and evaluated.

Mechanical Properties. Epoxy thermoplastics can be fabricated using mainstream plastics processes like extrusion, injection molding, blow molding and thermforming, and, because of their structural diversity, can exhibit widely different mechanical properties (5,6). Amorphous, glassy polymers, the most prevalent epoxy thermoplastics, typically have tensile or flexural moduli above 2.1 GPa (300 ksi), and while some of these materials are rather brittle (21,23), a number of epoxy thermoplastics have the ductility and impact resistance associated with engineering thermoplastics. Table V indicates that injection-molded samples of high-molecular-weight (Mw = 130,000) poly(hydroxy ether) **2a**, synthesized from bisphenol A and highly pure **1** (see above), is a strong (break strength = 53.1 MPa) and rather stiff (modulus = 2.1 GPa) thermoplastic that is, nonetheless, fairly ductile as reflected by a tensile elongation of 60% (16). In addition, this lightly branched **2a** is remarkably impact resistant with a notched Izod impact value of 1276 J/m, performance superior to that of polycarbonate, which is often utilized for its combination of stiffness and toughness. More importantly, poly(hydroxy amino ethers) with useful barrier properties also are mechanically quite robust. Ethanolamine-based **3a** (O_2TR = 0.8 BU; Tables I and IV) and piperazine-containing **3b** (O_2TR = 1.5 BU; see Table I for structure), both of which have high moduli (2.4 GPa), break at about 47 MPa and show good yielding behavior with elongations of 46 and 27 %, respectively (Table V) (9). With notched Izod impact values of 107 and 961 J/m, **3a,b** certainly have sufficient impact resistance for rigid or flexible packaging. Both polymers have good practical toughness; in Dynatup experiments, injection molded disks of the resins require over 50 ft-lb of energy for complete penetration and fail by a ductile rather than a brittle mechanism (9). Especially high-barrier poly(hydroxy amino ethers), such as **3h** (Table IV), have tensile properties similar to those of **3a**.

Table V. Mechanical Properties of Some Amorphous Epoxy Thermoplastics

Property:	Polymer 2a	Polymer 3a	Polymer 3b
Yield Stress, MPa (psi)[a]	59.3 (8600)	57.2 (8300)	55.6 (8060)
Break Stress, MPa (psi)[a]	53.1 (7700)	47.6 (6900)	47.4 (6870)
Elongation, %[a]	60	46	27
Flex Modulus, GPa (ksi)[b]	2.11 (305)	2.42 (351)	2.44 (354)
Notched Izod, J/m (ft-lb/in)[c]	1276 (24)	107 (2)	961 (18)

Properties were determined for injection molded specimens according to [a]ASTM D-638, [b]ASTM D-790, and [c]ASTM D-256.

The attractive mechanical performance of poly(amino ether) **3b** shown in Table V is retained in the semi-crystalline variant **3e** (x = 0.75; Tm = 213 °C), injection-molded samples of which yield at 56.4 MPa and break (with little post-yield stress drop) at 52.4 MPa. The polymer retains a rather high modulus (2.93 GPa) while exhibiting astonishingly high impact resistance with a notched Izod value of almost 1600 J/m (Table VI). Mechanical properties of liquid-crystalline poly(hydroxy ether) **2g** also are impressive. Conventionally configured, injection-molded ASTM tensile bars of **2g**, which have the classic skin-core composition expected for a thermotropic resin, display high yield and break stress near 88 MPa with 11% elongation, along with nylon-like toughness indicated by a notched Izod impact value of 316 J/m (Table VI). As would be anticipated for a liquid crystalline thermoplastic, higher strengths are realized when more uniform orientation of **2g** is achieved by injection molding into thin sections. Thus, in the machine direction, 40-mil plaques of the polymer have a yield and break stress of about 118 and 107 MPa, respectively (with no loss in elongation; Table VI), and break stresses as high as 130.9 MPa (18,980 psi) accompanied by moduli over 5.07 GPa (735 ksi) were measured for 10-mil micro tensile specimens of **2g** in which surface and core orientation are virtually identical (28). In sharp contrast to the behavior of high-modulus **2g** and **3e**, semi-crystalline poly(hydroxy ester ethers) such as **6d** (Table III) offer rubber-like toughness (no break in a notched Izod experiment) and high tensile elgonations (up to 357%); a more detailed description of **6d** and other poly(hydroxy ester ethers) will be presented in a forthcoming publication.

Table VI. Mechanical Properties of Some Crystalline Epoxy Thermoplastics

Property:	Polymer 3e[a]	Polymer 2g[b]	Polymer 2g[c]
Yield Stress, MPa (psi)[d]	56.4 (8180)	86.8 (12,590)	117.7 (17,060)
Break Stress, MPa (psi)[d]	52.4 (7600)	89.2 (12,940)	106.6 (15,454)
Elongation, %[d]	33	11	13
Tensile Modulus, GPa (ksi)[d]	2.93 (425)	3.2 (471)	3.70 (537)
Notched Izod, J/m (ft-lb/in)[d]	1593 (30)	316 (6)	-

[a]Values are shown for **3e** with x = 0.75 and Tm = 213 °C. [b]Standard injection-molded ASTM tensile bar. [c]Injection-molded plaque with a thickness of 40 mil. [d]See footnotes to Table V.

References

1. McAdams, L. V.; Gannon, J.A. In *Encyclopedia of Polymer Science;* Kroschwitz, J. L., Ed.; Wiley: New York, 1986, Vol. 6; 322.
2. Reinking, N. H.; Barnabeo, A. E.; Hale, W. F. *J. Appl. Poylm. Sci.* **1963**, *7*, 2135.
3. Reinking, N. H.; Barnabeo, A. E.; Hale, W. F. *J. Appl. Polym. Sci.* **1963**, *7*, 2145.
4. Hale, W. F., Phenoxy Resins. *Encyclopedia of Polymer Science and Technology;* Wiley: New York, 1969; Vol. 10, pp 111-122.
5. White, J. E.; Silvis, H. C.; Mang, M. N.; Brennan, D. J.; Schomaker, J. A.; Haag, A. P.; Kram, S. L.; Brown, C. N. *Polym. Prepr. (Am. Chem. Soc., Div. Polym. Chem.)* **1993**, *34* (1), 904.
6. White, J. E.; Silvis, H. C.; Brennan, D. J.; Mang, M. N.; Haag, A. P.; Schomaker, J. A.; Kram, S. L.; Brown, C. N. *SPI Epoxy Resin Formulators Division, Proceedings;* 1994; Chapter 3.
7. Salame, M.; Temple, E. J. *Adv. Chem.* **1974**, *135*, 61.
8. Salame, M. *J. Plast. Film Sheeting* **1986**, *2*, 321.
9. Silvis, H. C.; White, J. E. *Polymer News* **1998**, *23*, 6.
10. Silvis, H. C. *Trends Polym. Sci.* **1997**, *5 (3)*, 75.

146

11. Silvis, H. C.; White, J. E. (The Dow Chemical Company). U.S. Patent 5 275 853, 1994
12. Silvis, H. C.; Kram, S. L. (The Dow Chemical Company). U.S. Patent 5 464 924, 1995.
13. Flory, P. J. *Chem.Rev.* **1946**, *39*, 137.
14. Silvis, H. C.; White, J. E.; Crain, S. P. *J. Appl. Polm. Sci.* **1992**, *44*, 1751.
15. White, J. E., Ringer, J. W. (The Dow Chemical Company). U.S. Patent 5 164 472, 1992.
16. Schomaker, J. A.; White, J. E.; Haag, A. P.; Pham, H. Q. (The Dow Chemical Company). U.S. Patent 5 401 814, 1995.
17. White, J. E.; Brennan, D. J.; Pikulin, S. (The Dow Chemical Company). U.S. Patent 5 089 588, 1992.
18. Brennan, D. J.; Silvis, H. C.; White, J. E.; Brown, C. N. *Macromolecules* **1995**, *28*, 6694.
19. Brennan, D. J.; White, J. E.; Haag, A. P.; Kram, S. L.; Mang, M. N.; Pikulin, S.; Brown, C. N. *Macromolecules* **1996**, *29*, 3707.
20. Brennan, D. J.; Haag, A. P.; White, J. E.; Brown, C. N. *Macromolecules* **1998**, *31*, 2622.
21. White, J. E.; Haag, A. P.; Pews, R. G.; Kram, S. L.; Pawloski, C. E.; Brown, C. N. *J. Polym. Sci., A, Polym. Chem.* **1996**, *34*, 2967.
22. White, J. E.; Haag, A. P.; Pews, R. G. (The Dow Chemical Company). U.S. Patent 5 149 768, 1992.
23. Mang, M. N.; White, J. E.; Haag, A. P.; Kram, S. L.; Brown, C. N. *Polym. Prepr. (Am. Chem. Soc., Div. Polym. Chem.)* **1995**, *36* (2), 180.
24. Mang, M. N.; White, J. E. (The Dow Chemical Company). U.S. Patent 5 171 820, 1992.
25. Mang. M. N.; White, J. E.; Swanson, P. E. (The Dow Chemical Company). U.S. Patent 5 496 910, 1996.
26. White, J. E.; Kram, S. L.; Brown, C. N.; Bicerano, J.; Unpublished results.
27. Brennan, D. J.; White, J. E.; Brown, C. N. *Macromolecules* (in press).
28. White, J. E.; Silvis, H. C.; Mang, M. N.; Kram, S. L.; Hefner, Jr., R. E. (The Dow Chemical Company). U.S. Patent 5 686 551, 1997.
29. Navard, P.; Haudin, J.-M. In *Polymeric Liquid Crystals*; Blumstein, A., Ed., Penum: New York , pp 389-398.
30. DeLassus, P. Barrier Polymers. *Encyclopedia of Chemical Technology*, 4th ed.; Wiley: New Your, 1992; Vol. 3, pp 931-962.

Chapter 11

Allyl Alcohol- and Allyl Alkoxylate-Based Polymers

Shao-Hua Guo

Lyondell Chemical Company, 3801 West Chester Pike, Newtown Square, PA 19073

Allyl alcohol and allyl alkoxylate are incorporated into a variety of polymers to provide the pendant hydroxyl group for crosslinking reaction. The polymers are prepared by a bulk free radical polymerization process in which the allyl monomer is charged into the reactor before the polymerization starts, and the vinyl comonomers and the initiator are gradually fed into the reactor at the polymerization temperature. The polymers are oligomers of molecular weight in a range of 500 to 10,000. The hydroxyl functional polymers derived from allylic alcohols are useful materials in coatings, adhesives, elastomers, and may other thermoset polymers. This article reviews the chemistry and uses of allylic monomers. The free radical polymerization of allylic monomers is described in detail. Several examples of commercially usefull resins prepared from allylic monomers are provided.

Allyl alcohol and its derivatives have been widely employed as intermediates in chemical synthesis and as monomers in polymerization. Although allyl compounds exist widely in nature, they are prepared synthetically for large commercial volumes. Early syntheses of allyl alcohol was based on the preparation of allyl halide intermediates from glycerol or propylene. There are two more recent commercial routes to produce allyl alcohol: 1) oxidizing propylene to acrolein and then reacting it with a secondary alcohol, and 2) isomerizing propylene oxide. The route 1) results in a mixture of allyl alcohol and a ketone. The route 2) is preferred to produce a high purity allyl alcohol.

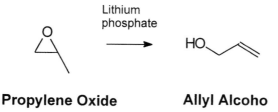

Lithium
phosphate

Propylene Oxide   **Allyl Alcoho**

Most of the allyl alcohol produced is consumed in the manufactures of glycerol and 1,4-butanediol. Allyl alcohol is oxidized with hydrogen peroxide in an aqueous stream to yield glycerol. Hydroformylation of allyl alcohol produces 1,4-butanediol and 2-methyl-1,3-propanediol.

Because of their low polymerization reactivity, mono-allyl monomers have found limited commercial use in polymers. However, several difunctional or mutilfunctional allyl compounds have found wide applications in thermosetting polymers [1]. These include diallyl phthalate (DAP), allyl diglycol carbonate, allyl acrylate or methacrylate, and many others. DAP can be prepared by direct esterification of phthalic anhydride using an excess of allyl alcohol [2], by transesterification of dimethyl phthalate with allyl alcohol in the presence of sodium methylate [3], or by the reaction of phthalic anhydride with allyl chloride in the presence of an equivalent amount of sodium carbonate as a reactant [4]. DAP offers advantages over styrene in unsaturated polyester applications including less odor and volatility, more stable syrups in storage, and higher heat-resistance of the cured resins.
However, high cost and slow curing have limited its uses.

Diallyl Phthalate (DAP)

Allyl diglycol carbonate is prepared by the reaction of diethylene glycol bis(chloroformate) with allyl alcohol in the presence of alkali [1]. Like diallyl phthalate or other allyl monomers, allyl diglycol carbonate has very good storage stability even mixed with a peroxide initiator. Allyl diglycol carbonate is polymerized by a bulk free radical polymerization to produce cast sheets, lenses, and other shapes of products. The products exhibit outstanding scratch resistance, optical properties, and impact strength.

Allyl methacrylate or acrylate can be prepared by alcoholysis of methacrylate acrylate by allyl alcohol in the presence of sodium methylate and a polymerization inhibitor such as hydroquinone [5]. The greater reactivity of the acrylic double bond leads to the formation of a soluble copolymer containing allyl groups that allow a second stage polymerization to form crosslinked thermoset plastics and

ADGC

coatings. Up to 10% of allyl methacrylate or acrylate may be copolymerized with methyl methacrylate or other acrylate comonomers in a broad applications including dental plastics, lenses, adhesives, fiber-reinforced plastics, acrylic ester rubbers, printing inks and coatings [6]. The allyl acrylic monomers have been also used as crosslinking agents along with polyesters, diallyl esters , styrene, and acrylonitrile.

Allyl alcohol readily reacts with an alkylene oxide to form an allyl alkoxylate, for example, allyl propoxylate or allyl ethoxylate [7]. Allyl alkoxylates have considerably lower acute toxicity and lower vapor pressure than allyl alcohol, and are easier to handle in the preparation of polymers [8]. Allyl alcohol and allyl alkoxylates undergo free radical copolymerization with vinyl comonomers such as vinyl aromatics, acrylates and methacrylates, vinyl ethers and esters, vinyl halides, and conjugated dienes. The allyl monomers offer hydroxyl functionality to the copolymers. Due to the low reactivity of the allyl double bonds, the copolymerization can be carried out in a semi-batch bulk polymerization process in which the vinyl comonomer is gradually fed into the allyl monomer. The allyl monomer regulates the molecular weight of the copolymer owing to its low chain propagation rate and strong chain transfer reaction. The copolymers usually have a molecular weight in a range of 500 to 50,000 depending on the percentage of the allyl monomer in the copolymer. In a free radical copolymerization, the allyl monomer has a reactivity ratio which is close to zero, and , therefore, it offers an even distribution of hydroxyl group along a polymer chain. Three styrene and allyl alcohol copolymers are commercial available from Lyondell Chemical Company. Hydroxyl acrylic resins based on allyl alcohol and allyl alkoxylate are under development.

RESULTS AND DISCUSSION

Allyl propoxylate monomer can be prepared by reacting allyl alcohol with propylene oxide in the presence of a basic catalyst. The reaction yields a mixture of allyl propoxylates which have various chain length of propoxylation. The average number n of oxypropylene units can be controlled by the ratio of allyl alcohol and propylene oxide. Allyl monopropoxylate is easily separated from a mixture of propoxylates since it has a boiling point 50°C below the allyl dipropoxylate. A short chain allyl propoxylate (n<2) is preferred for polymerization because it is relatively easy to remove after polymerization by distillation. An increase of every one oxypropylene unit increases the boiling point of the allyl propoxylate about 50°C. For example, allyl monopropoxylate and allyl dipropoxylate have boiling points of 145°C and 195°C, respectively. Like other allyl monomers, allyl propoxylate monomer is not completely polymerized in a free radical polymerization, thus the unreacted monomer must be removed from the product. The high boiling point of the long chain propoxylate (n>2) increases the difficulty of removing and recycling the unreacted monomer. Listed in Table 1 are the physical properties of allyl alcohol and allyl propoxylate.

Table 1   Physical Properties of Allyl Alcohol and Allyl Propoxylate

Monomer	Allyl Alcohol	Allyl Monopropoxylate	Allyl Propoxylate With Average 1.6 Oxypropylene Units
Molecular Weight	58.1	116.2	150
Boiling Point, °C	97.1	145	145-245
Freezing Point, °C	-129	not tested	not tested
Flash Point (TCC), °C	21.1	not tested	63
Homopolymer T_g, °C	4	-15	-33
LD 50, Oral, mg/Kg	65	not tested	1100

Allyl Alcohol And Allyl Propoxylate Homopolymers and Copolymers [9]
Homopolymers of allyl alcohol and allyl propoxylate have been prepared by free radical polymerization. It has been found that allyl propoxylate is considerably more reactive than allyl alcohol. As showed in Table 2, the homopolymerization of allyl propoxylate yields 75% of polymer, while allyl alcohol gives only 28% of polymer under similar polymerization conditions. In the copolymerization of allyl alcohol and allyl propoxylate, the polymer yield increases with the decrease of the ratio of allyl alcohol to allyl propoxylate.

Poly(allyl propoxylate) is soluble in all of the commonly used solvents listed in the Table 3. Poly(allyl alcohol) is insoluble in most solvents but alcohols. The solubility of the copolymers of allyl alcohol and allyl propoxylate increases with increase of the allyl propoxylate units. The difficulty of preparation of poly(allyl alcohol) and its poor solubility have limited its application, however, poly(allyl propoxylate) has showed increased potential as a highly hydroxyl-functionalized polymer.

Table 2 Allyl Alcohol and Allyl Propoxylate Homopolymers and Copolymers

Polymer	Composition (in Mole)	Mn (GPC)	Mw/Mn (GPC)	wt% Yield of Polymer
A	Poly(Allyl Alcohol)	-*	-*	28
B	Poly(Allyl Monopropoxylate)	1080	1.6	75
C	90/10 Allyl Alcohol and Allyl Monopropoxylate Copolymer	580	2.0	53
D	50/50 Allyl Alcohol and Allyl Monopropoxylate Copolymer	1000	3.0	56

* Not tested because poly(allyl alcohol) in insoluble in THF.

Table 3 Solubility of Allyl Alcohol and Allyl Propoxylate Homopolymers and Copolymers

Solvent	Polymer A	Polymer B	Polymer C	Polymer D
Methanol	Soluble	Soluble	Soluble	Soluble
Propylene Glycol t-butyl ether	Insoluble	Soluble	Partially Soluble	Soluble
Tetrahydrofurane	Insoluble	Soluble	Partially Soluble	Soluble
Acetone	Insoluble	Soluble	Partially Soluble	Partially Soluble
Methyl Ethyl Ketone	Insoluble	Soluble	Insoluble	Insoluble
Ethyl Acetate	Insoluble	Soluble	Insoluble	Insoluble
Xylene	Insoluble	Soluble	Insoluble	Insoluble

Copolymers of allyl alcohol and allyl propoxylate with styrene [10]   Allyl
alcohol and allyl alkoxylate undergo free radical copolymerization with styrene.
Three styrene-allyl alcohol copolymers (SAA Resins) are commercially available.
They are widely used in polyurethane, melamine, alkyd, and many other coatings
and thermoset polymer systems. The aromatic character of the SAA resins
improves the hydrolytic stability of derivatives and increases the resistance of the
polymers containing SAA to water, detergents, chemicals and corrosion [11]. The
basic difference among various SAA resins is in the copolymer composition, i.e.,
the monomeric unit ratio of styrene to allyl alcohol. The molecular weight of SAA
resins is determined by its composition. The higher the allyl alcohol monomeric
concentration, the lower the molecular weight. This is resulted from the
combination of the low chain propagation rate and the strong chain transfer of the
allyl alcohol monomer.

Table 4 Compositions and Physical Properties of SAA Resins

Polymer	SAA-103	SAA-100	SAA-101
Composition Styrene/Allyl Alcohol Molar Ratio	20 : 80	30 : 70	40 : 60
Molecular Weight Mw, GPC	8400	3000	2500
Molecular Weight Mn, GPC	3200	1500	1200
Hydroxyl Number mgKOH/g	125	210	255
Glass Transition Temperature,°C	78	62	57

Copolymers of allyl propoxylate and styrene have similar characteristics to
the SAA resins. Allyl propoxylate is of particular interest in preparing water-
soluble or water-dispersible polymers which contain acid functional groups. These
polymers are prepared by copolymerizing allyl propoxylate, styrene, and acrylic
acid or methacrylic acid [12]. Listed in the Table 5 is the composition and physical
properties of a water reducible copolymer of allyl propoxylate, styrene and
methacrylic acid. The terpolymer is soluble in the commonly used organic solvents,
and is water soluble or dispersible after the acid functional groups are neutralized.
Organic amines such as triethylamine, trimethylamine, diethanolamine,
triethanolamine, N,N-dimethylethanolamine are preferred neutralization agents,
although alkali hydroxides, alkaline hydroxides and ammonia can be also used.
        Like SAA resins, the water-soluble or water-dispersible terpolymer of allyl
propoxylate, styrene and methacrylic acid can be used in water borne polyurethan,
melamine, polyester, and many other water borne coatings and thermoset polymers
to improve performance.

Table 5 Composition and Physical Properties of Water Reducible Copolymer

Copolymer Composition, wt%	
Allyl Monopropoxylate	26
Styrene	68
Methacrylic Acid	6
Total	**100**
Physical Properties	
Molecular Weight, Mn	2500
Molecular Weight, Mw	5500
Hydroxyl Number, mgKOH/g	125
Acid Number, mgKOH/g	70
Glass Transition Temperature, oC	65
Solution Viscosity	
Methanol	
Ethanol	
Ethyl Acetate	
Butyl Acetate	
Methyl Iso-butyl Ketone	
Methyl Amyl Ketone	
Propylene Glycol Methyl Ether Acetate	
Propylene Glycol Methyl Ether	
Ethylene Glycol Methyl Ether	
Xylene	

Hydroxyl Acrylic Resins [13]   Allyl alcohol and allyl alkoxylate have been incorporated into hydroxyl acrylic resins to provide the pendant hydroxyl groups for crosslinking reaction. The resins are crosslinked with melamine or an isocyanate to form acrylic-melamine coatings or acrylic-polyurethane coatings, respectively. The coatings can be used as automotive topcoat or other high performance industrial coatings.

The existing hydroxyl acrylic resins are copolymers of a hydroxyl functional acrylate or methacrylate, one or more of ordinary acrylate or methacrylate such as methyl methacrylate, butyl methacrylate and butyl acrylate, and optionally styrene. The hydroxyl functional acrylate and methacrylate commonly used include hydroxyethyl acrylate or methacrylate, and hydroxypropyl acrylate or methacrylate. The resins are made by a free radical polymerization in the presence of a chain transfer agent or solvent. The polymerization is usually conducted at the reflux temperature of the solvent in the range of 120°C to 150°C. The high polymerization temperature and the use of the chain transfer solvent lower the molecular weight of the acrylic resin to the range from 1,000 to 5,000. The low molecular weight of the resin results in low VOC (Volatile Organic Compound) of the final coating formulation.

Unlike the hydroxyalkyl acrylates or methacrylates, which have close monomeric reactivity ratios to the comonomers, allyl alcohol and allyl alkoxylate are considerably lower in reactivity than the acrylate and styrene comonomers. A large excess of the allyl monomer is needed to achieve the desired incorporation of allyl monomeric unit into the copolymer, and the unreacted allyl monomer is removed after polymerization. To minimize the amount of allyl monomer removal and recycle, a semi-batch polymerization process is employed in which all of the allyl alcohol or allyl propoxylate monomer needed for the batch production is charged into the reaction vessel along with a small part of the total amount of acrylate comonomers and the initiator before the polymerization starts, and the remaining acrylate comonomers and initiator are fed into the reaction vessel during polymerization in decreasing rates. The polymerization is carried out usually for 4 to 8 hours at a relatively high temperature, 125°C to 165°C. The high polymerization temperature is required to achieve a high polymerization rate and high polymer yield. The total monomer conversion can be up to 95% depending on the copolymer composition to be made and the polymerization process condition. No solvent is needed in the polymerization.

When allyl alcohol is used, a pressure reaction vessel is needed because the polymerization is usually conducted at temperature which is above the boiling point, 97oC, of allyl alcohol. Allyl monopropoxylate and monoethoxylate have boiling points of 145°C and 150°C, respectively. The polymerization involving those monomers can be carried out at the reflux temperature of the allyl monomer under an atmosphere pressure.

To determine the hydroxyl functionality distribution of the copolymer against molecular weight, an acrylic copolymer has been fractionated to 10 fractions by the supercritical CO_2 technique by Phasex Corperation, and each fraction is analyzed by GPC for molecular weight and ^{13}C-NMR for composition. As showed in Table 6, the copolymer composition is dependent on the molecular weight. The lower the molecular weight, the higher concentration of allyl propoxylate unit. The dependence of the molecular weight on the copolymer composition is very consistent with the results of the copolymer of methyl methacrylate and allyl alcohol studied under low-conversion polymerization with various comonomer feed composition [14]. This distribution of hydroxyl functional group against molecular weight is very desirable for application in coatings and other thermoset polymers because the low molecular weight fractions of high hydroxyl content can effectively participate crosslinking reactions.

Table 6 Fractionation of the Copolymer of Methyl Methacrylate and Allyl Propoxylate

Fraction#	Mass of Fraction, g	Mn, By GPC	Mole % of Methyl Methacrylate Monomeric Unit	Mole % of Allyl Propoxylate Monomeric Unit
1	9.00	750	59.1	40.9
2	4.04	1070	63.1	36.9
3	6.70	3670	69.5	30.5
4	6.58	5490	75.3	24.7
5	6.93	5840	79.7	20.3
6	2.93	10670	82.7	17.3
7	3.68	16240	85.3	14.7
9	5.18	23940	87.7	12.3
9	7.65	35790	89.8	10.2
10	6.38	58170	93.8	6.2
Parent		3403	79.0	21.0

EXPERIMENTAL

Safety Allyl alcohol is highly toxic and flammable liquid. It is corrosive to skin and is a severe eye irritant. It is also a high inhalation, skin absorption, and ingestion hazard. It can cause damage to the liver and kidneys, cause lacrimation, tremors, excitability, convulsion, and even coma. Allyl propoxylate has dramatically reduced toxicity. It is a severe eye irritant, and is a slight ingestion hazard. Study the Material Safety Data Sheets of the monomers and have an adequate personal protection before handling allyl alcohol.

Materials   Allyl monopropoxylate and allyl propoxylate with 1.6 of an average oxypropylene unit are development products from Lyondell Chemical Company. Other materials are available from Aldrich. They are used as received.

Homopolymerization of Allyl Alcohol   A one liter stainless steel reactor equipped with a mechanical agitator, steam heating jacket, temperature controller, addition pump, and inlets for nitrogen and vacuum, is charged with 432 grams of allyl alcohol. The reactor is purged with nitrogen and sealed. After heating to 150°C, 61 grams of di-tert-butylperoxide is continuously added to the reactor through the addition pump over 4 hours. Heating is continued at 150°C for 0.5 hour after the initiator addition is complete. The reactor content is cooled to 25°C, and the unreacted allyl alcohol is removed via vacuum distillation (1.5 mmHg) by slowly heating to 165°C. The product of allyl alcohol homopolymer is collected after cooling to 25°C.

Homopolymerization of Allyl Propoxylate   The procedure is the same as the Homopolymerization of Allyl Alcohol except allyl propoxylate instead of allyl alcohol is used.

Copolymerization of Allyl Alcohol and Allyl Propoxylate   The procedure is the same as the Homopolymerization of Allyl Alcohol except a mixture of allyl alcohol and allyl propoxylate is used. The comonomer feed composition of allyl alcohol and allyl propoxylate is varied to achieve copolymers of different compositions.

Copolymerization of Styrene and Allyl Alcohol For Preparing SAA-100   500 grams of allyl alcohol and 16 grams of di-t-butylperoxide is charged into the one liter stainless steal reactor as in Homopolymerization of Allyl Alcohol. The reactor is purged with nitrogen and sealed. 234 grams of styrene and 8 grams of di-tere-butylperoxide are mixed and chilled to 5°C to avoid polymerization of the mixture. The mixture is loaded into the addition pump. The reactor is heated to 145°C, and the mixture of styrene and di-tere-butylperoxide is pumped into the reactor at the temperature reading over a period of 7 hours in a decreasing rate. The addition rates are as follows: 50 grams per hour for the first hour, 45 for the second hour, 40 for the third hour, 35 for the fourth hour, 30 for the fifth hour, 24 for the sixth hour, and 18 for the seventh hour. Heating continues for 0.5 hour at 145°C after the addition is complete. The reactor is cooled to 25°C. The excess allyl alcohol and a small amount of unreacted styrene are removed from the polymer product via vacuum distillation (1.5 mmHg) by slowly heating up to 165°C.

Various SAA copolymers are made via a similar procedure by changing the ratio of the total amount of allyl alcohol and the total amount of styrene.

Copolymerization of Styrene and Allyl Propoxylate   The procedure is the same as in Copolymerization of Allyl Alcohol and Styrene except allyl propoxylate instead of allyl alcohol is used. Various styrene and allyl propoxylate copolymers can be made by changing the ratio of styrene to allyl propoxylate.

Copolymerization of Allyl Propoxylate with Methyl Methacrylate   The procedure is the same as in Copolymerization of Styrene and Allyl Propoxylate except methyl methacrylate instead of styrene is used. Various copolymers of allyl propoxylate and methyl methacrylate can be made by changing the ratio of methyl methacrylate to allyl propoxylate.

Copolymerization of Allyl propoxylate, Styrene and Methacrylic Acid   500 grams of allyl monopropoxylate is charged into a one liter stainless steal reactor as in Homopolymerization of Allyl Alcohol. 34 grams of T-hydro 70 (70% aqueous solution of tert-butyl hydroperoxide), 50 grams of methacrylic acid, and 200 grams of styrene are mixed and chilled to 5°C to avoid polymerization of the mixture. 90 grams of the mixture is charged into the reactor and the rest into the addition pump. The reactor is purged with nitrogen and sealed. After the reactor is heated to 145°C, the mixture of styrene, methacrylic acid and T-hydro 70 in the addition pump is added into the reactor over a period of 5 hours in a decreasing rate. The addition rates are 58 grams per hour for the first hour, 48.5 for the second hour, 39 for the third hour, 29 for the fourth hour, and 19.5 for the fifth hour. Heating continues for 0.5 hour at 145°C after the addition is complete.   The reactor is cooled to 25°C. The unreacted monomers are removed from the polymer product via vacuum distillation (1.5 mmHg) by slowly heating up to 165°C.

CONCLUSION

Allyl alcohol and allyl alkoxylate are useful hydroxyl functional monomers. These monomers undergo free radical copolymerization with most commonly used vinyl monomers to form oligomers. Allyl alcohol and allyl alkoxylate not only offer the copolymer hydroxyl functionality, but also control molecular weight of the copolymers and regulate the polymerization rate so that the polymerization does not need a solvent or chain transfer agent. Many of the hydroxyl functional copolymers are of potential applications in coatings, adhesives, elastomers, and other thermoset polymers. A high hydroxyl content is found in the low molecular weight fraction of the copolymer. This unique feature allows the copolymers of very low molecular weight to effectively participate crosslinking reaction.

ACKNOWLEDGEMENT

This work was supported by the Acrylic Polyol Project team of Lyondell Chemical Company. Mark Smithson, Robert Good, Julia Weathers, David Pangburn, and Val Pearson conducted most of the experimental work. David Kinney provided NMR analysis.

REFERENCES

1.   C. E. Schildknecht, *Kirk-Othmer* Encyclopedia of Chemical Technology, 3rd ed., Wiley-Interscience Publishers, New York, NY, **1978**, Vol 2, p. 97.

2.   *U.S. Patent 2,311,327,* T.F. Bradley *(to Shell).*

3.   U.S. Patent 2,218,439, Herry S. Pothrock (To E. I. du Pont de Nemours & Company).

4.   U.S. Patent 3,250,801, Hugo Stange, and William B. Tuemmler, (to FMC Corporation)

5.   S. G. Cohe, *J. Polym. Sci.* **1948**, 3, 278.

6.   C.E. Schildknecht, *Polymerization Processes, High Polymers,*, Interscience Publishers, a division of John Wiley & Sons, Inc., New York, **1977**, Vol 29, Chapt. 2.

7.   I. Swern, *J. Am. Chem. Soc.*, **1949**, *71,* 1152.

8.   S. Harris, A. Good, R. Good, and S. H. Guo, *Modern Paint and Coatings*, 1994, November, 34.

9.   U. S. Patent 5,451,652, S. H. Guo and R. G. Gastinger (To ARCO Chemical Company).

10.   U. S. Patent 5,382,642, S. H. Guo (To ARCO Chemical Company).

11.   D. Pourreau, S. H. Guo and B. J. Corujo, *J. American Paint & Coatings,* **1995**, *May,* 43.

12.   U. S. Patent 5,728,777, S. H. Guo (To ARCO Chemical Company)

13.   U. S. Patent 5,646,213, S. H. Guo (To ARCO Chemical Company).

14.   S. H. Guo, *Solvent Free Polymerizations and Processes, ACS Series Book*, 713, Chapter 7, p113-126, **1998**

Chapter 12

New Monomers from Vegetable Oils

E. H. Brister, T. Johnston, C. L. King, and S. F. Thames

Department of Polymer Science, The University of Southern Mississippi,
Hattiesburg, MS 39406

Novel vegetable oil derived macromonomers were synthesized by the chemical modification of castor, lesquerella and vernonia oils. The castor oil derivative has been successfully copolymerized into vinyl-acrylic and all-acrylic latexes designed for use as architectural coatings. The castor oil derived monomer provides latex polymers with good film formation without the use of coalescing solvent. Furthermore, its presence in the copolymer improves dry times, and provides reactive functionality for ambient cure. In yet another phase of the investigation of novel monomers, UV curable coatings were developed from methacrylic macromonomers synthesized from lesquerella and vernonia oils. The resulting coatings exhibited good gloss, pencil hardness, and adhesion to steel substrates.

Coatings are essential to our society in that they play a major role as protective barriers to the destructive forces of the environment, and they provide aesthetic value. However, traditional disadvantages of many coating types center around their release of volatile organic compounds (VOCs), and thus they often possess strong odor. The polymer component of a coating formulation provides its integrity, and acts as the matrix, or binder. However, it is the polymer component of a formulated coating that requires solvents. Solvents either solubilize the polymers as in organic soluble polyester compositions or plasticize latex particles to ensure coalescence in traditional latex technology. In either case, VOC emissions and strong odor are the end results. In an effort to meet our environmental concerns with high performance products, the principles of vegetable oil, air dry coatings have been adopted and adapted to the synthesis of novel latex polymers.

For instance, novel macromonomers have been designed, synthesized and utilized in the preparation of emulsion polymers for coatings applications. Specifically, the latex polymers do not require traditional cosolvents for film formation, and they possess little to no odor.

In summary, vegetable oils provide a renewable resource for development of polymers for use in coatings applications. The vegetable oils are triglyceride esters of fatty acids, and vary in chain length and functionality, making them a versatile starting material (*1*). Vegetable oil derived macromonomers for coatings applications are obtained by chemical modification of the triglyceride, thereby allowing application in various types of coatings including architectural and UV curable coatings.

Castor Oil Macromonomer

Castor oil contains approximately 85% ricinoleic acid, a C_{18} hydroxy fatty acid which is characterized by the presence of an isolated double bond and a secondary hydroxyl group. The Castor Acrylated Monomer (**CAM**) is synthesized by first converting castor oil to its methyl ester, i.e. methyl ricinoleate, and then acrylating the remaining hydroxyl group (Scheme 1). The resulting multifunctional, acrylic monomer is uniquely structured to provide latex polymer synthesis via the acrylate moiety, and crosslinking after application via oxidative polymerization of the residual double bond. CAM is a yellow liquid which blends well with vinyl acetate and acrylic monomers, and thereby can be utilized in common polymerization processes. CAM has a molecular weight of 366 grams/mole, making it one of the largest, yet reactive, comonomers used in emulsion polymerization to date. It has been successfully copolymerized into acrylic and vinyl-acrylic copolymer latexes, and thus was chosen as a model compound for investigating the structure-property relationships of latex polymers.

Architectural coatings are approximately 50% of the total coatings market (*2*). This class of coatings is applied and develop final film properties at ambient conditions. Acrylic and vinyl-acrylic monomers (*3*) are principal articles of commerce for the architectural coatings market, yet they do not crosslink. The principal mechanism of film formation for an acrylic or vinyl acrylic latex is via cosolvent plasticized latex particle coalescence, with concurrent water and coalescent solvent evaporation. Coalescent solvent(s) remain in the polymeric film for the longest periods, and during evaporation contributes significantly to VOCs and odor.

Acrylic latexes are known for outstanding exterior performance, while vinyl-acrylic latexes are known as the workhorse for interior architectural coatings because of their cost-performance and versatility (*4*). To be effective architectural binders, latex polymers must form a rigid, continuous protective barrier over the substrate. The formation of latex films relies on polymer chain entanglement for integrity and performance rather than chemical crosslinks. The paradox between the rigidity required for performance and the flexibility necessary for film formation creates a dilemma, which is typically resolved by addition of volatile coalescing solvents in order to plasticize the latex polymer thereby enhancing diffusion across particle boundaries. Hard monomers, or monomers with high T_g homopolymers, are judiciously balanced with soft, flexibilizing comonomers such as butyl acrylate, 2-ethylhexyl acrylate, and vinyl versatate to generate latex polymers with a T_g appropriate for architectural coatings. Added coalescing solvents enhance molecular diffusion and hence continuous film formation of latexes with T_gs above ambient. The coalescing solvents are fugitive plasticizers which gradually volatilize, contributing to odor and volatile organic content (VOC) in addition to increasing the toxicity, flammability and cost of the latex.

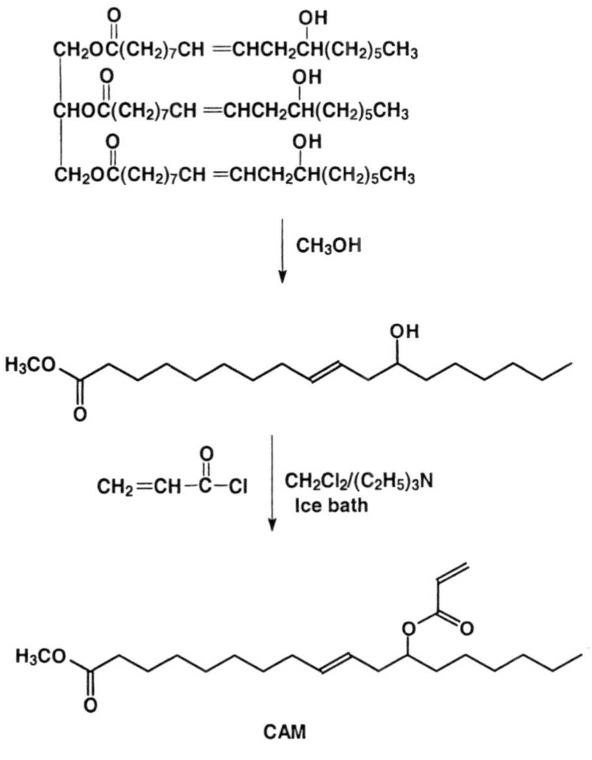

Scheme 1. Synthesis of CAM (reproduced with permission from ref. 6. copyright 1998)

The development of vegetable oil derived monomers is based on the desire to combine the well established auto-oxidative curing mechanism of oil modified polyesters with the properties of conventional latexes, while eliminating coalescing solvent requirements for the latter. For instance, in oil modified polyesters, the auto-oxidation process of vegetable oils leads to molecular weight enhancements, T_g increases, improved solvent and chemical resistance, and enhanced durability. Consequently, oil modified polyesters have set the standard for cost-performance in architectural coatings, and represent the single largest quantity of solvent-soluble resins used in coatings (5). Thus, the concept of utilizing vegetable oil derived monomers in latex polymer compositions allows the coating scientist to incorporate the well known auto-oxidative crosslinking functionality with the added benefit of internal plasticization and low to no odor.

In order to exploit the concept of oxidative drying of latexes, little to no VOCs and low odor, CAM has been utilized in the design and synthesis of emulsion polymers. The data obtained confirm that CAM is a powerful coalescing aid, and essentially eliminates the need for coalescing solvents (6).

However, the production of CAM containing acrylic and vinyl-acrylic latexes require alternative synthetic approaches in order to minimize chain transfer interference. Thus, the preparation of CAM containing vinyl-acrylic and acrylic latexes are described separately, due to their significant differences in reactivity and hydrophobic character.

Vinyl-Acrylic Latexes Containing CAM. A synthetic challenge for the synthesis of vinyl-acrylic latexes is to minimize chain transfer and termination reactions due to the highly reactive vinyl acetate radicals (7). Chain transfer to polymer is a frequent occurrence and is influenced by processing parameters in the emulsion polymerization of vinyl acetate (8). In addition to chain transfer, grafting and chain branching occur in vinyl-acrylic latexes when stabilizing colloids are employed as part of the surfactant system, particularly in the presence of acrylic comonomers (9). An additional chain transfer mechanism can occur in vinyl-acrylic latexes using unsaturated vegetable oil derived comonomers as a result of labile allylic hydrogens which provide a mechanism for chain transfer to monomer.

A surfactant system based exclusively on anionic and nonionic surfactants was developed for vinyl-acrylic latex synthesis using CAM concentrations as high as 15 parts per hundred monomer (phm) by weight. In an effort to optimize the level of CAM in emulsion polymers, a study of gel formation as a function of CAM concentration was undertaken. CAM vinyl-acrylic latex conversions were shown to decrease, while the quantity of insoluble material (gel) increased with increasing CAM concentration (Table I). The decrease in conversion is likely a result of chain transfer to CAM. The high gel content accompanied by low conversion is a consequence exacerbated by undesirable branching and grafting producing crosslinked, insoluble structures. An advance in polymerization technique utilizing a staged monomer addition protocol reduced the chain transfer to monomer and allowed the synthesis of CAM vinyl-acrylic latexes with conversion comparable to that of controls synthesized without CAM. This technique produced emulsion polymers with a very low quantity of insoluble material.

CAM incorporation in the vinyl-acrylic latex polymers was verified using NMR spectroscopy (10). Polymer composition was determined by ^{1}H NMR analysis, and CAM incorporation was evidenced in both ^{13}C and ^{1}H NMR spectra. The polymer composition indicates that CAM is incorporated in vinyl-acrylic latex polymers even more efficiently than butyl acrylate. NMR data also indicate that some finite amount of the CAM isolated double bonds are retained throughout the emulsion polymerization, thereby allowing for post polymerization/auto-oxidative cure.

Acrylic Latexes Containing CAM. The ability of CAM to function as a coalescing aid for acrylic latexes was demonstrated when compared to the plasticizing effect of a commercially common cosolvent, Texanol™. Atomic force microscopy (AFM) in

Table I: Vinyl-acrylic Latex Data

Latex	CAM phm	Feeding protocol	% Conversion	% Gel (40 mesh)
VAc 1	0	Conventional	99	0
VAc 2	5	Conventional	92	1
VAc 3	15	Conventional	81	12
VAc 4	5	Staged	99	0
VAc 5	15	Staged	99	2

tapping mode was used to contrast the surface topology of an acrylic latex film containing 4 phm CAM versus a lab control, possessing the same T_g, plasticized by 10 g/L Texanol (Figure 1). The CAM acrylic latex exhibited a more completely coalesced surface compared to the control with added cosolvent, as indicated by discernable distinct particles.

Chain transfer in acrylic polymerizations was measured by the amount of gel in the latex products (Table II) (11). Reaction temperature, monomer feed time, and percent CAM have all been found to influence gel content. No gel was observed in surfactant stabilized acrylic latexes with CAM concentrations at or less than 1 phm, regardless of feed time or polymerization temperature. With CAM concentrations between 5 and 10 phm, the acrylic polymerizations had more than 60% gel when the reaction temperatures were above 65°C, and a monomer feed time of two hours. Gel was also observed to increase in 5 phm CAM latexes when the feed time was increased to greater than three hours. Latexes containing higher gel contents produce marginal particle coalescence resulting in poor scrub resistance. Conversely, low polymerization temperatures (such as 45°C) with moderate amounts of CAM (5-10 phm) and short monomer feed times (less than three hours), resulted in latexes that had complete coalescence and exhibited good scrub resistance.

Lesquerella Oil Macromonomers

The major fatty acid component of lesquerella oil (LO) is lesquerolic acid (55%), a C_{20} hydroxy fatty acid with an isolated double bond. Other than castor oil lesquerella is the only commercially available oil with a hydroxyl functional fatty acid. Two types of LO methacrylates, methacrylated lesquerella oil (MALO) and hydroxyethyl methacrylate modified lesquerella oil (HEMALO), were synthesized by reacting the oil with methacryloyl chloride and an methacrylic functional isocyanate, respectively (Scheme 2).

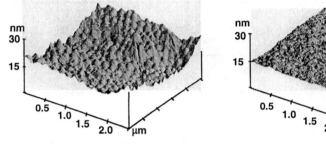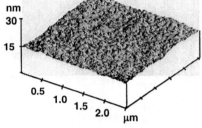

Acrylic latex control with 10 phr Texanol   Acrylic latex with 4 phr CAM

Figure 1. Acrylic latex film formation depicted by AFM micrographs (reproduced with permission from ref. 6. copyright 1998)

Table II: Acrylic Latex Data

Latex	CAM phm	Reaction Temp. (°C)	% Conversion	% Gel (250 mesh)
AA1	0	45	99	0
AA2	1	45	99	0
AA3	5	45	100	0
AA4	10	45	99	0
AA5	15	45	99	16
AA6	0	65	96	0
AA7	1	65	99	0
AA8	5	65	96	62
AA9	10	65	99	64
AA10	0	80	100	3
AA11	1	80	99	3
AA12	5	80	99	65
AA13	10	80	100	75

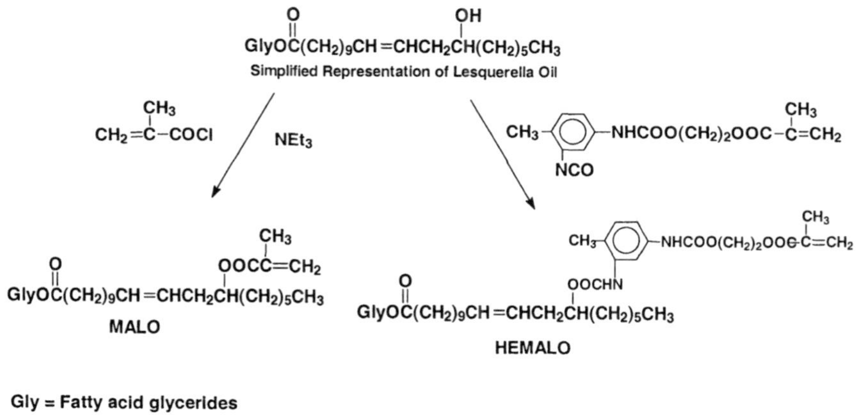

O
‖
GlyOC(CH₂)₉CH=CHCH₂CH(CH₂)₅CH₃
$$GlyOC(CH_2)_9CH=CHCH_2\overset{\overset{\displaystyle OH}{|}}{C}H(CH_2)_5CH_3$$

Simplified Representation of Lesquerella Oil

Scheme 2. Synthesis of LO acrylates (reproduced with permission from ref. 14. copyright 1998)

Ultraviolet Cured Coatings Using LO Methacrylates.

Ultraviolet (UV) curing technology provides substantial energy savings, shorter production times, and faster cure rates. UV coatings can be easily applied with conventional equipment onto a variety of substrates including plastics and heat sensitive materials. Limitations thus far have been due to optimization of the raw materials (12-13) and interference from light absorbing pigments.

LO methacrylates possess low toxicity and no detectible volatility (14-15). Coating formulations utilizing LO methacrylates are given in Table III, and coating characteristics are included in Table IV. LO methacrylate derived coatings exhibit inherently low viscosities at 100% solids, and are devoid of hazardous emissions during UV curing. VOC free, 100% solids coatings were applied over wood, aluminum, and steel substrates. The film properties of LO methacrylic coatings were characterized by excellent gloss and pencil hardness values as compared to a control polymer. Direct and reverse impact strength values over steel substrates were better than the control, while the crosshatch adhesion results were comparable.

Vernonia Oil Macromonomer

Vernonia oil (VO), in its natural state, contains epoxide functionality within its fatty acid chains. The epoxide functionality provides reactive sites for derivation to polymerizable groups. For instance, methacrylic derivatives of VO (MVO) were synthesized by reacting VO with methacrylic acid in the presence of a tertiary amine (Scheme 3) (15). The extent of epoxide methacrylation was determined to be 65% using ¹H NMR analysis (16). MVO possessed lower viscosity than traditional

Table III. UV Curable Coatings Formulations

Materials	Coating 1 (Control)	Coating 2	Coating 3
HEMALO,g	-	15.1	-
MALO,g	-	-	15.0
Commercial Photomers,g	72.2	73.0	72.4
BYK 065, g	0.4	0.4	0.4
BYK 325, g	0.5	0.5	0.5
Irgacure 651, g	2.1	2.5	2.5
Benzophenone, g	1.1	1.3	1.3

Table IV. UV Cured Coatings Properties

Property	Coating 1			Coating 2			Coating 3		
	Wood	Al	Steel	Wood	Al	Steel	Wood	Al	Steel
Gloss (20°)	85	95	93	81	100	72	76	84	79
Yellowness index		1.3			2.3			3.0	
Pencil Hardness	9H	F	8H	9H	4H	7H	6H	5H	7H
Direct Impact, J			4.5			9.1			7.9
Reverse Impact, J			0.6			2.3			2.3
Adhesion	5B	0B	1B	5B	3B	4B	5B	0B	5B

methacrylates prepared from epoxidized soybean oil (*16*) (ESO), and thus offers improved processing and formulation advantages. The viscosity of the VO based coating is 321 cps while the viscosity of an identically formulated coating from ESO is 918 cps.

The methacrylic ester polymerized rapidly in the presence of other UV active comonomers (Table V). The cured coating had good pencil hardness (>3H), high MEK rub resistance (>300 rubs), and good adhesion.

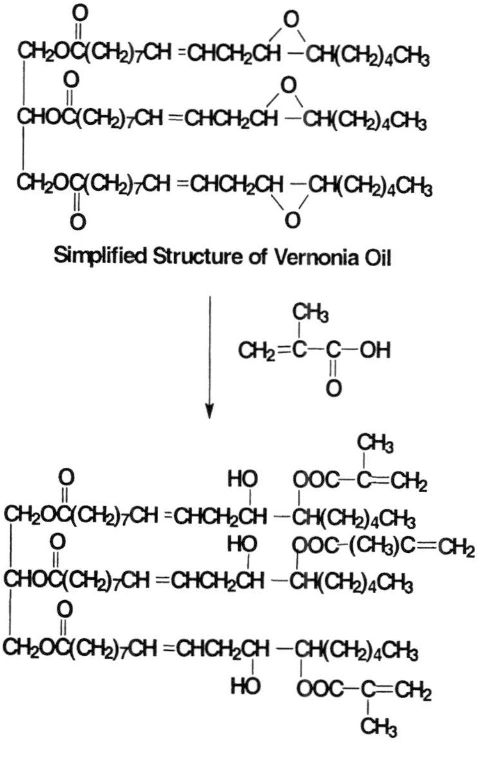

Simplified Structure of Vernonia Oil

Scheme 3. Synthesis of VO methacrylates (reproduced with permission from ref. 14. copyright 1998)

Table V: UV Curable Coating Formulation

Component	MVO Formulation	ESO Formulation
MVO	14.0	-
ESO	-	14.0
Photomer 4127	12.0	12.0
Photomer 4061	10.0	10.0
Photomer 4094	15.0	15.0
Photomer 4149	3.9	3.9
Photomer 4770	5.0	5.0
Photomer 3016	12.4	12.4
Photomer 6008	8.0	8.0
Silwet 7002	0.4	0.4
Byk 065	0.3	0.3
Byk 325	0.4	0.4
Irgacure 651	1.0	1.0
Photomer 81	0.8	0.8

Conclusions

Vegetable oil derived macromonomers were successfully synthesized from castor, lesquerella, and vernonia oils. CAM was effectively incorporated in vinyl-acrylic and all-acrylic latexes and facilitated film formation in the absence of coalescing solvents. Lesquerella and vernonia methacrylate macromonomers were synthesized and formulated into environmentally friendly UV cured coatings. The lesquerella and vernonia modified coatings exhibited good gloss, pencil hardness, and adhesion to steel substrates.

Acknowledgments

The authors would like to acknowledge Dr. Akhtar Khan and Dr. Ramesh Subramanian for their help and support.

Literature Cited

1. Eckey, W. *Vegetable Fats and Oils,* Reinhold Pub, NY 1954.
2. Bourguignon, E. W. *Paint & Coatings Industry,* **1998**, 14(2), 32.
3. Kline Industry Survey Report, *Synthetic Latex Polymers,* **1998**.
4. Klein, R. J. *Modern Paint and Coatings,* Mar **1993**, pp 37-39.
5. *Surface Coatings,* Oil and Colour Chemists' Association, Chapman Hall, NY, **1983**, Vol. I, pp 53.
6. Thames, S. F. *Proc. 25th Int. Waterborne, High-Solids, and Powder Coat. Symp.,* **1998**, pp 305-320.
7. Blackley, D. C. *Emulsion Polymerization,* John Wiley & Sons, N.Y., **1975**, pp 442.
8. Britton, D; Heatley, F; and Lovell, P. A. *Macromolecules,* **1998**, *31,* 2828-2837.
9. Craig, D. H. In "Water Soluble Polymers, Beauty with Performance," Glass, J. E. Ed.; Advances in Chemistry Series, American Chemical Society, **1986**, *213,* 351-367.
10. Brister, E. H.; Smith, O. W.; and Thames, S. F. *Proc. 26th Int. Waterborne, High-Solids, and Powder Coat. Symp.,* February **1999**.
11. King, C. L.; Smith, O. W.; and Thames, S. F. *Proc. 26th Int. Waterborne, High-Solids, and Powder Coat. Symp.,* February **1999**.
12. Sawyer, R *Modern Paint and Coatings,* **1991**, No.(6), pp 34.
13. Guarino, J. P. *Modern Paint and Coatings,* **1991**, No.(6), pp 38.
14. Thames, S. F.; Yu, H; Wang, D. M.; and Schuman, T. P. Low- and No-VOC Coating Technologies, 2nd *Biennial International Conference,* **1995,** Durham, NC.
15. Thames, S. F.; Blanton, M. D.; Mendon, S.; Subramanian, R.; and Yu, H. In *Surfactants and Fatty Acids: Plant Oil in Biopolymers from Renewable Resources,* Ed. Kaplan, D. L.;, Springer-Verlag: Heidelberg, New York, **1998,** Ch. 10; pp 249-280.
16. Johnston, T.; and Thames, S. F. The University of Southern Mississippi, unpublished results.
17. Smith, O.W.; Borden, G.W.; and Trecker, D.J. US Patents 4233130, 4224369, 4220569, 4215167, 4157947, 4025477, 4016059, 3979270, 3931071, 3931075, 3878077, and 3876518.

Chapter 13

Surface Modification of Silica Particles and Silica Glass Beads

R. M. Ottenbrite[1], R. Yin[1], H. Zengin[1], K. Suzuki[1], and J. A. Siddiqui[2]

[1]Department of Chemistry, Virginia Commonwealth University, Richmond, VA 23284–2006
[2]DuPont Polyester Films, P.O. Box 411, Hopewell, VA 23860

The surface properties of particles are very important with respect to their end use. Many particles are used for fillers and to impart properties that enhance a materials function. In order to do so, the particles need to be compatible with the environment into which they are placed. Often this is in a polymer matrix such as a film. To enhance particle compatibility of the surface of 2-5 μm silica particles and silica glass beads, surface modification of these particles was carried out. Surface modifications were accomplished by surface polymerization, surface polymer grafting, by surface dendrimerization and by developing organo-silicone particles. The surfaces obtained can have a wide variety of properties, from highly hydrophilic to highly hydrophobic, from anionic and cationic to nonionic, as well as being environmentally responsive.

Interest in organic and inorganic composite materials appears to be constantly increasing due to their superior properties. Inorganic materials, for example, are frequently used as fillers or coatings in many polymer systems. Generally the surface properties are most important in determining the quality of the finished material[1]. It is known, for example, that a solid formed from a liquid can engulf or reject fine dispersed particles depending on the surface energy of the particles and the matrix[2]. Particle rejection from the solid would clearly lead to a very inhomogeneous distribution of the particles in a filler application. Similarly, using particles as coatings would largely determine the surface properties of the coated object. In order to obtain predictable performance from coatings and fillers, it is desirable to create well defined surfaces that are also thoroughly characterized.

The importance of silica particles, for example, has been recognized for a variety of applications. Different types of silica, including porous silica and nonporous glass beads, have been developed and applied in environmental science, chemical processing, mineral recovery, energy production, agriculture and pharmaceutical industries. Grafting reactions[3], on porous silica[3-5] or nonporous glass beads[6] have also attracted attention and have been studied extensively. Surface modification of silica particles and glass beads, for example, has been studied in our group to determine grafting reaction conditions such as time, temperature, initiator, monomer and glass beads concentration. An analysis of these grafting parameters under different reaction conditions indicates new grafting mechanisms[7].

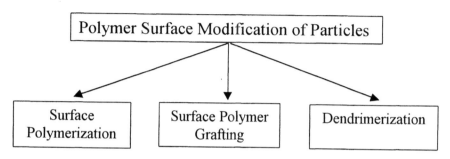

Composites containing micron-sized particles are probably even more promising for applications such as chromatography, chemical sensors, ceramics and biocompatible materials. In these cases, particles are generally surface modified. Due to potential heterogeneity of solid surfaces, it is desirable to have flexibility in building complex surface layers. Considering these needs, the dendritic modification of glass beads has been carried out in our laboratory[8,9]. In those experiments, the conditions for efficient chemical surface modification were studied to achieve highly functional and well controlled surfaces.

Polymer Grafting of Silica Glass Beads

Polymer grafting of nonporous glass beads (2-5nm), which can be used as fillers in polymer matrices to enhance polymer properties, has the potential for unique new materials. Compared with porous silica, which has an abundance of silanol groups on the surface as well as a larger total surface area, the glass beads that were used in this study have a very low silanol concentration (about 0.21mmol/g) and a small surface area ($\sim$2m^2/g). These surface property shortcomings, however, can be overcome by surface modification via polymer grafting[7]. Potentially, a selective polymer surface layer on the glass beads can improve their compatibility and adhesion with a polymer matrix. We studied the effects of the grafting reaction conditions such as reaction time, temperature, initiator, monomer and the total bead surface area that can affect the graft polymerization.

Generally, there are two approaches to grafting a polymer chain onto a surface[7]. (a) via free radical copolymerization of an immobilized double bond by using a single coupling agent such as methacrylpropylsiloxane or vinyl trimethoxylsilane, and (b) via initiation by an immobilized free radical intiator[5]. Using these two methods, polymers have been grafted onto silica or porous silica surface[5] and only a limited amount of data on nonporous substrates has been reported[6,10].

We pretreated P_2 glass beads with sodium hydroxide to expose more hydroxyl groups to the coupling agent methacrylpropylsiloxane (MPS). The MPS modified beads were then allowed to react with acrylic acid and benzoyl peroxide (BPO) or 2,2-azobisisobutyronitrile (AIBN) to provide polymer-grafted beads. Generally, AIBN is not considered to be a suitable grafting intiator because of the resonance stabilization of the insipient radical[11]. However, AIBN was as effective as BPO at producing poly(acrylic acid) grafts on glass surfaces from dioxane solutions. This graft polymerization process produced both grafted polymer and free polymer. The free polymer was separated from the polymer modified beads by Soxhlet extraction with a 1:1 methanol/acetone mixture. The solvent mixture effectively separated the free homopolymer from the grafted polymer; Tsuzuki et al[12] found that the extraction with only methanol gave a higher apparent grafting percentage than true grafting.

The molecular weight was determined by gel permeation chromatography on a μ-Styragel 10^3 column with water as the mobile phase and poly(styrene sulfonate) was used to calibrate the column. Although there is some difference between the molecular weight of the graft polymer and homopolymer[13], the concept that the two polymers should have relatively similar molecular weights is commonly accepted[13-15].

The formation of polymer grafts on the bead surface occurred either by addition of macroradical homopolymers to the double bonds on the methacrylpropylsiloxane (MPS) modified glass surface or by the primary radical initiation of these double bonds to polymerize acrylic acid. Higher grafting efficiencies and grafting percentages were obtained in dioxane in comparison with tetrahydrofuran (THF). Isopropanol, a chain transfer solvent, produced a lower molecular weight graft and grafting percentage. Benzene and toluene were not found to be good solvents because of their poor solubility for the free polymer and the grafted beads tended to form agglomerates in these solvents.

The AIBN system grafted polymers to the glass surface by the addition of macroradicals to the vinyl groups on the surface, therefore, space hindrance played an important role in amount of grafted polymer obtained. In the case of BPO, primary radicals also seem to be produced at the glass surface by radical addition to the immobilized vinyl group, which then react with the acrylic acid monomers to form the graft polymer. BPO appeared to be the better overall initiator for this graft process[16].

An increase in the temperature accelerated the decomposition rate of the initiator and molecular diffusion. Consequently the grafting percentage was greater at higher temperature. Increasing the initiator concentration also caused an increase in the number of grafts and a decrease in molecular weight of the graft polymer. Decreases in the grafting percentage at high initiator concentrations were ascribed to the decrease in molecular weight of the graft polymer.

Immobilization of Carboxylic Acid Polymers on P_2 Glass

In this study polymer immobilization involved the reaction of carboxylic acid polymers (CAPs) with APS (3-aminopropyl siloxane) functionalized glass beads (Scheme 1). The polymers, PAA and PMA reacted with the active APS amine groups to form salts.

Upon heating, these salts formed stable amide bonds between the APS modified beads and the polymers. The carboxyl and amine functional groups react chemically at 140°C to form a covalent bond[17, 18].

The effect of PAA concentration on polymer grafting was examined and a plateau was observed at 10 mg PAA/mL in DMF which was considered to be the optimal polymer concentration. The effects of PAA molecular weight on the modification of glass beads were also studied, and it was found that the higher molecular weight polymers gave only slightly higher percent grafted polymer[17].

The dispersability of PAA modified glass beads composites were examined in different solvents. In polar solvents, such as methanol and acetonitrile, the particles showed good dispersability (Table 1). In non-polar solvents, such as aliphatic and aromatic hydrocarbons, the particles did not disperse very well. The dispersability of the composite particles relates to the hydrophilic/hydrophobic compatibility between PAA on the glass surface and the solvent.

Table 1 **The dispersibility of the polyacrylate salt-glass beads composites in different solvents**

Solvents	Dispersibility	Solvents	Dispersibility
Methanol	Excellent	Petroether	None
DMF	Very Good	Benzene	None
DMSO	Very Good	Pentane	None
Acetonitrile	Very Good	Dioxane	None
NH₄OH	Excellent	Cyclohexane	None
Ethylene Glycol	Excellent	Toluene	None
Water	Very Good	THF	Poor

The presence of PAA polymer on the glass surface before heating was confirmed by the appearance of bands at 1701.6 cm^{-1} and 1560.6 cm^{-1} due to C=O stretching of the carboxylic acid and the free carboxylate groups of the polymer, respectively. The IR data indicate that the polymer coating on glass exists in two states, the carboxylic acid and the carboxylate ion[18]. The lower IR frequency bands of the polymer overlapped with the strong and broad band due to Si-O stretching. The bands at 3448.1 cm^{-1} and 1654.6 cm^{-1} are due to O-H bond stretching and deformation. It is difficult to determine whether any silanol groups had reacted with the polymer based on the very weak characteristic peak at 3753.6 cm^{-1}.

The APS modified beads treated with PAA produced new peaks at 1710 cm^{-1} and 1660 cm^{-1}, indicating that amide bonds had formed between the carboxyl groups of the PAA and the amine group of bound APS, respectively. A small band at 1700-1705 cm^{-1} was attributed to the carboxylic group (COOH →1700-1705 cm^{-1}), while the band at 1672 cm^{-1} corresponds to the amide group (Figure 1).

To determine the percentage weight versus temperature, thermogravimetric analysis (TGA), was carried out with 20 mg samples of modified glass beads at a heating rate of 10°C/min in a dynamic air and helium mixture. The onset of the decomposition for all thermograms was ca. 270°C and the weight losses are shown in their respective thermograms (Figure 2). Thermogram A depicts the changes, which

Figure 1. IR Spectra of polymer-glass composite at different reaction times.

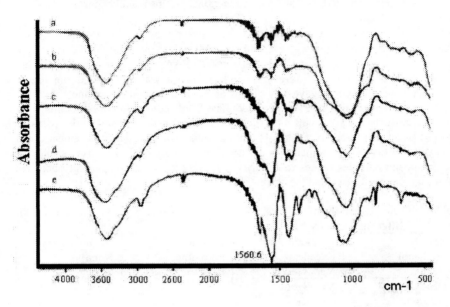

Figure 2. Thermogravimetric Curves of (A) Glass Beads (B) APS, and (C) PAA.

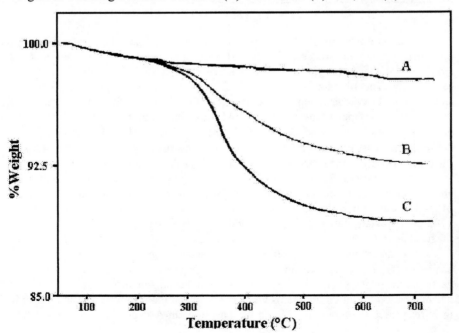

are minimal for the glass beads alone. Thermogram B shows a weight loss of 10 wt % after grafting PAA. Based on this work, other poly(carboxylic acid) polymers and anhydride copolymers, such as poly(styrene-co-maleic anhydride) have been to the surface of P_2 beads as well[19].

Modification of Porous Silica Particle Surfaces with Poly(Acrylic Acid)

The importance of silica particles has been recognized for a variety of applications. Different types of silica including porous silica and nonporous glass beads have been developed and applied in environmental science, chemical processes, mineral recovery, energy production, agriculture and pharmaceutical industries. Grafting on porous silica[5, 20-21] has attracted much attention and has been studied extensively.

The present study is concerned with the modification of functional polymers onto porous silica particle surfaces. Our primary interest is to improve particle surface characteristics. Poly(acrylic acid) was chosen as the functional polymer to provide pH-intelligent, surface-responsive particles. The PAA chains under acid conditions are usually coiled, while under basic conditions the chains are extended due to electrostatic repulsion of the carboxylate ions. By controlling the pH, the surface characteristics can be tailored to respond to specific pH environments. Pore size distribution and specific surface area of modified silica are calculated from the amount of nitrogen adsorbed on the surface. The water penetration rate and porosity for different pH were measured for estimation of the surface properties[6-7, 22].

Porous silica particles were obtained from Suzuki Oil & Fat Co. Ltd. with particle size distribution of 2.0-2.5 µm. The silica particles were treated with a 2% sodium hydroxide solution, washed with water and dried.

Silica surface was treated with APS and modified with poly(acrylic acid), using the reaction illustrated in Scheme 1. The carboxyl groups on poly(acrylic acid) reacted with the APS amine functional groups to form salt bonds on the surface of silica. The salt-modified beads were heated in DMF at 140°C to form amide bonds. The FTIR spectrum of the APS-modified silica exhibited a peak at about 1600 cm^{-1} due to the amino group in addition to the original silica peak. Treatment with poly(acrylic acid) produced new peaks at 1710 cm^{-1} and 1660 cm^{-1}, indicating the carboxyl groups of poly(acrylic acid) and the amide bond formation between the poly(acrylic acid) and the bound APS, respectively[23].

It was found that the surface modification of silica with APS and poly(acrylic acid) decreased the total pore volume. The specific surface area also decreased with each modification step. Therefore, by controlling the amount of poly(acrylic acid) on the surface, it is possible to control the total pore volume. Although no significant change could be found for water penetration rate for the APS-modified silica, the rate of water uptake by the poly(acrylic acid) coated particle changed with the pH. Similarly, the porosity of APS-modified silica was not influenced by pH, but the porosity of poly(acrylic acid)-modified silica changed with pH. The poly(acrylic acid) chains were extended or coiled accordingly to the electrostatic interactions between charged groups which are sensitive to pH and the ionic strength of the aqueous solution. Dissociation of poly(acrylic acid) at pH 5 extended the chains, while at pH 1.5 the chains remain closer to the surface. These observations are in agreement with

Scheme 1   Reaction for modification of APS beads with PAA by grafting

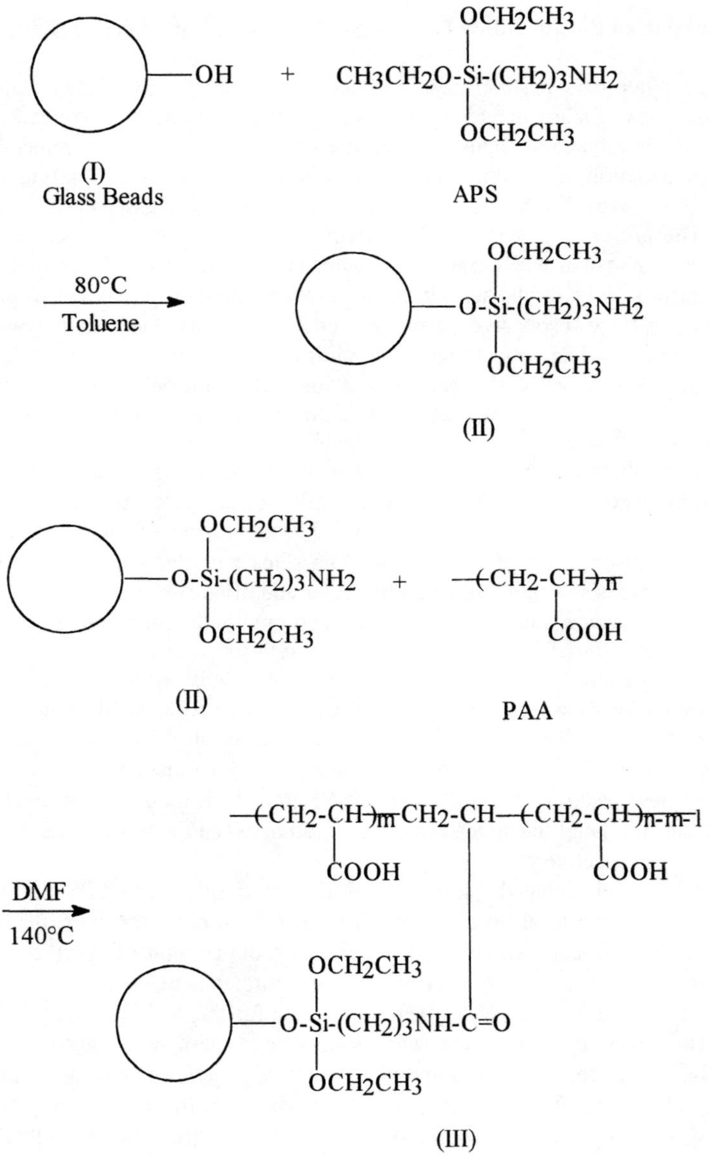

(I)
Glass Beads

APS

(II)

(II)

PAA

(III)

experiments from Ito et al.[24], who reported that poly(acrylic acid)-grafted porous polycarbonate membrane showed higher permeabilities at lower pH than at higher pH.

The amount of grafted polymer increased with increasing polymer concentration. The highest amount of PAA grafted on silica was 396 mg/g-silica[28]. These polymer modified silica particles are being characterized and evaluated for potential film applications.

Dendritic Modification of Glass Bead Surfaces

A new type of glass bead modification, which is based on the design of divergent dendrimer structures and fractals[24-27], is also being pursued in our laboratory. We pursued this method to compensate for the low silanol (=SiOH) concentration and small surface area of the glass beads compared to the abundance of silanol groups and larger surface area of porous silica. We visualized the glass beads as a starburst core, therefore, the surface functional groups on the glass bead could be bonded with multifunctional molecules to develop fractal structures on the glass surface. The dendritic procedure provides the possibility to place a great number of small molecules and reactive groups on the glass bead surface. Using this technique, the properties of the glass beads could be controlled to form hydrophobic or hydrophilic surfaces.

Dendritic modifications were carried out by first etching the glass beads and coupling with APS (aminopropyltrimethoxysilane) to form the first generation. Propanolamine was reacted with the residual methoxyl groups to introduce more amine groups on the glass bead surface, forming the second generation. Next, trifunctional cyanuric chloride or alkyl chloride was reacted with amine groups to build the third generation. Finally, a variety of groups, which offered different physical properties such as propanolamine, tris(hydroxylmethyl) aminomethane, or methyl tyrosine, were grafted to develop a fourth generation. The definition of the different generations is given in Table 2. The grafting percentage and conversion were calculated from thermogravimetric (TGA) analysis based on the weight loss of the modified glass beads during temperature increases. The temperature was scanned from 50-600°C at a rate of 2.5°C/min.

Table 2 Reaction conversions at different generations

Generations	Reaction Conversions (%)		
	GF-1-2-3-2*	GF-1-2-3-4*	GF-1-2-3-5*
1st	55.2	44.6	24.6
2nd	5.59	8.23	55.1
3rd	92.9	63.2	9.52
4th	38.2	52.1	5.66

*GF: glass bead; the digits after FG represent the chemicals which reacted with glass to form different generations. 1-APS; 2-Propanolamine; 3-Cyanuric chloride; 4-Tris; 5-Methyl tyrosine.

After the coupling reaction, IR spectra of every sample demonstrated a broad band at 1578.2 cm^{-1} due to N-H bonding vibration. This indicated that the APS was grafted on the glass surface to provide amine groups. When the second generation was formed on the glass surface, the IR band at 1578.2 cm^{-1} was enhanced. According to the analytical data, one propanolamine was successfully added per APS to increase the number of surface amine groups. After cyanuric chloride was reacted with the amine groups, a new band at 1613.5 cm^{-1} appeared in the IR spectra, which was ascribed to C=N stretching vibration. Finally, three different molecules with different functional groups were attached, respectively, to provide the last generation, which were identified by a band at 2902.2 cm^{-1} due to C-H stretching vibration of the CH$_2$ groups of propanolamine and Tris and at 1689 cm^{-1} due to carbonyl vibration for −C(NHR)O. These data were supported by the results from TGA and titrametric analyses. The fourth generation of GF-1-2-6 and GF-1-2-7 consisted of long alkyl chains of palmitoyl chloride and lauroyl chloride, respectively. Therefore, the IR spectra show strong C-H vibrations at 2902.2 cm^{-1} due to CH$_2$ group. IR spectra of the different generations for GF-1-2-3-4 are illustrated in Figure 3.

The grafting percentage increased with dendritic growth, as shown in Figure 4. Therefore, this dendritic procedure increased the number of grafted molecules on the glass bead surface[8]. The highest w/w grafting percentage obtained by this procedure was 21%. The same grafting percentage was also achieved by polymer coating on the glass bead surface[28]. However, the dendritic procedure bonds numerous small molecules that cover the glass surface with a uniform coating.

It is difficult to obtain 100% conversion for any generation since the percent conversion changed alternatively at each generation. This means that the fourth generation would give a low conversion if a high conversion appeared in the third generation and vice versa. This phenomenon may be characteristic of the dendritic growth process. A reasonable explanation is that space hindrance plays an important role in the dendritic growth. Another factor affecting the third generation is that cyanuric chloride could react with more than one amine group in the same generation. As a result, the number of reactive points of the next generation would be decreased. These two factors could explain the conversion changes for GF-1-2-3-5 at different generations. In the second generation, the conversion was 55% and then decreased to a very low value (9.5%) for the next generation. In general, when one generation has a low grafting percentage, a higher conversion could be achieved for the next generation. These results were based on normal reaction conditions; i.e. the reagents reacted were about 10% in excess of the reactive sites. Exhaustive reaction conditions such as 10-fold excess of reagents and long reaction times were not employed.

The hydrophobic and hydrophilic properties of dendritical modified glass beads were evaluated in different solvents for dispersibility. GF-1-2-3-2 with a low hydrophilic end group exhibited a poor dispersibility in all solvents. However, GF-1-2-3-4 with a higher hydrophilic chain had a good dispersibility in water and methanol. For GF-1-2-3-5, which contains ester groups in the last generation, polar DMF is a good solvent. The last generations of GF-1-2-6 and GF-1-2-7 which were formed by long alkyl chains, show a good dispersibility in nonpolar solvents such as benzene and pentane. Therefore, the dispersibility of the modified glass beads is dependent on the molecular structure of the last generation on the modified glass bead surface[29].

Figure 3. Dendritic Modification of glass bead.

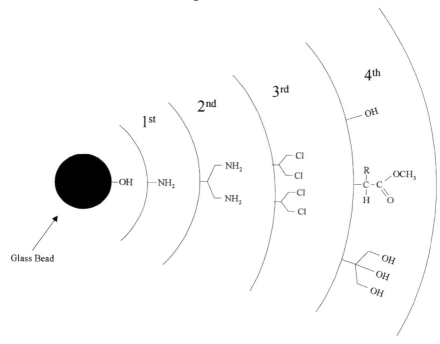

Figure 4. Grafting percentage increasing with dendritic growth

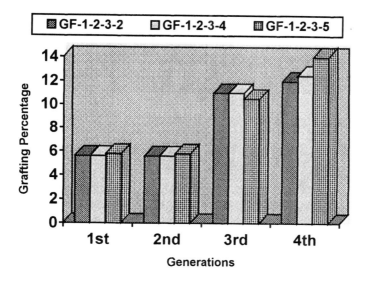

Alkylsilane Particles

A number of organosiloxane polymer systems utilized to prepare films by aerogel methods have been reported; however, the synthesis of controlled 'organic-inorganic' hybrid particles by the sol-gel process is still a challenging problem. Most of the sol-gel studies have focused on tetraethyl orthosilicate (TEOS) or tetramethyl-orthosilicate (TMOS) systems. However, by introducing organic groups into the inorganic network, many novel physical and chemical properties can be developed such as: a) improved compatibility with an organic matrix, b) specific surface properties, and c) changeable particle density. In general, organic-inorganic hybrid materials would include both the physical mixing and molecular level mixing. However, these physical mixtures are not stable to solvent and to aging. The organosiloxane hybrid particles would contain a homogeneous network of covalently bonded organic and inorganic structures.

A molecular level mixed-hybrid particle has been prepared by a one-step procedure in our laboratory[30]. An organic domain in hybrid particles was achieved by adding methyltriethoxysilane (MTEOS) to an alcohol/water/ammonia solution and shaking for 24 hours. Fine particles (2-3μm) were obtained with narrow polydispersity. Using a similar procedure with chloromethylphenylsiloxane (COPTOS), we were unsuccessful in producing stable particles.

Comparing COPTOS to MTEOS, methyltriethoxysilane has a methyl group attached to silicone which could influence the kinetics of the hydrolysis and condensation reactions. CMPTOS, however, contains an electron-withdrawing chloromethylphenyl group to stabilize the negative charge forming on silicon in the transition state. Based on the S_N2^--Si or S_N2^*-Si hydrolysis mechanism under basic catalyzed conditions, the silicon acquires a formal negative charge in the transition state. The methyl substituent, an electron donating group, should inhibit the formation of the negative charged, which would cause the hydrolysis rate to decrease. Therefore, MTEOS should undergo slower hydrolysis and condensation rates than CMPTOS which has an electron withdrawing group.

In general, if the hydrolysis rate is higher than the condensation rate, then complete hydrolysis will occur. However, if the hydrolysis rate is equal to or smaller than the condensation rate, the hydrolysis will be followed immediately by condensation. In the second case, some unhydrolyzed alkoxide groups will remain in the, thus the MTEOS derived particles may have some residual unhydrolyzed ethoxide groups trapped internally. According to the TGA data, the hydrolysis of MTEOS was completed in 8 hours in contrast to a higher hydrolysis rate for CMPTOS.

Based on the SEM (Figure 5), the particles are regular shape and exhibit a highly porous surface. Density analysis showed that the MTEOS particles were only 1.31 g/cm^3, which is much lower than 2.5 g/cm^3 for glass beads. The particle is an aggregate consisting of many smaller particles which are about 10 nm measured by particle size analyzer. The lower hydrolysis and condensation rate of MTEOS allowed us to observe particle growth, which is consistent with Iler's particle growth mechanism[31]. According to Iler's hypothesis, after nucleation, the particles grow from the nuclei by internal condensation of the hydroxyl groups within the particles. In contrast, the agglomerates derived from CMPTOS were less porous and spherical in shape which indicates that a different mechanism for particle formation took place.

Neither the TGA (Figure 6) nor IR analysis detected any surface water on the particles prepared from MTEOS, indicating a hydrophobic surface. After calcination at 700°C for 10 min, which pyrolyze the methyl group, the surface of the particles changed from hydrophobic to hydrophilic character[32]. The enhanced hydrophilic character was evident by an observed increase in the particle dispersibility in water and the appearance of absorbed H_2O on the particle surface by IR. The IR peak area at 1119cm^{-1} (O-C bond) decreased as the temperature changed from 180° to 250°C; simultaneously, the (Si-O) band at 1034^{-1}increased. This means that the unhydrolyzed alkoxide groups were being lost during the heating period. Above 400°C, the methyl groups gradually pyrolyzed resulting in a weaker adsorption at 1274 cm^{-1}.

The surface modification of polymethylsiloxane particles involved free radical chlorination. Carbon tetrachloride was employed to eliminate possible chain transfer by a solvent and small amounts of AIBN added to the reaction resulted in higher conversion[33].

The IR spectrum of the methylpolysiloxane particles, after chlorination, has a new band in the fingerprint region at 650cm^{-1} due to the C-Cl vibration. The next step was to substitute the chloride group with an amine group. A two step method was attempted for the chloride substitution using sodium azide and reducing the RN$_3$ with LiAlH$_4$ to the corresponding amine in ether. A better method to substitute an amine for the chloride group was with sodium amide in liquid ammonium. Higher reaction conversions were obtained due to better solubility of NaNH$_2$ in liquid ammonium than in other solvents tried. The presence of the amine was confirmed by IR and elemental analysis.

Further surface modifications of the amino alkylsilane particles included reactions; a) with succinic anhydride in THF to produce carboxylic groups, b) with isophorone diisocyanate in pyridine to provide the isocyanate group, a monomer reagent for polyurethane, c) with dimethyl 2,6-naphthalyl dicarboxylate in xylene to prepare an aromatic ester, and d) with cyanuric chloride to achieve dendritic modifications.

Acknowledgements

We wish to thank ICI Americas and High Technology Materials Center at Virginia Commonwealth University for their financial support for this project.

Figure 5. SEM pictures of polymethylsiloxane particles NH_3/Si: 28.9; (a) H_2O/Si: 58; (b) H_2O/Si: 5.5

a   b

Figure 6. TGA traces of particles prepared from(a) Tetraethoxysilane; and (b) Methyltriethoxysilane

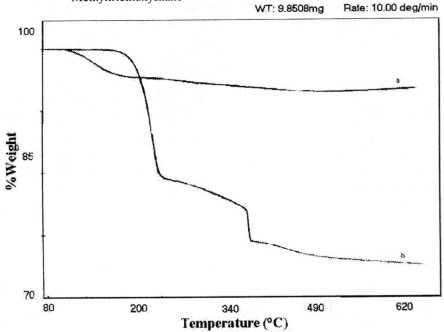

References

1. E.P. Plueddermann, Silane Coupling Agent; Plenum Press: New York, 1991, Chapter 1 & 2.
2. Li D.; Neumann A.W., Behavior of Particles at Solidifation Fronts, Applied Surface Thermodynamics, Marcel Dekker: New York, 1996, Chapter 12.
3. Carlier E.; Guyot A.; Revillon A., React. Polym., **1991**, 16, 115.
4. Nakatsuka T., J. Appl. Polym. Sci., **1987**, 34, 2125.
5. Boven G.; Oosterling M.L.C.M.; Challa G.; Schouten A.J., Polymer, **1990**, 31, 2377.
6. Boven G.; Folkersma R.; Challa G.; Schouten A.J., Polym. Commun., **1991**, 32, 50.
7. Yin R.; Ottenbrite R.M.; Siddiqui J.A., Polym. Prepr. **1994**, 35(2), 705
8. Yin R.; Ottenbrite R.M.; Siddiqui J.A., Polym. Prepr., **1995**, 36(1), 449.
9. Yin R.; Ottenbrite R.M.; Siddiqui J.A., Polym. Prepr., **1996**, 37(2), 751.
10. Hashimoto K.; Fujisawa T.; Kobayashi M.; Yosomiya R.J., Macromol. Sci., Chem., **1982**, A18, 173.
11. Nishioka N.; Matsumoto K.; Kosai K., Polym. J., **1983**, 15(2), 153
12. Tsuzuki M.; Hagiwara I.; Shiraishi N.; Yokota T., J. Appl. Polym. Sci., 1980, 25, 2909.
13. Nishioka N.; Kosai K., Polym. J., **1981**, 13(2), 1125
14. North A.M., The International Encyclopedia of Physical Chemistry and Chemical Physics, 1996, vol. 17(1), The Kinetics of Free Radial Polymerization, Pergamon Press.
15. Nishioka N.; Minami K.; Kosai K., Polym. J., **1983**, 15(8), 591
16. Yin R.; Ottenbrite R.M.; Siddiqui J.A., Polym. Adv. Tech., **1997**, 8, 761.
17. Ottenbrite R.M.; Zengin H.; Siddiqui J.A., Polym. Prepr., **1998**, 39(2), 585.
18. Ottenbrite R.M.; Zengin H.; Siddiqui J.A., Polym. Prepr., **1998**, 39(2), 587.
19. Zengin Huseyin, Masters Thesis, Virginia Commonwealth University, **1998**, May 17.
20. Naka T., J. Appl. Polm. Sci., **1987**, 34, 2125.
21. Carlier E.; Guyot A; Revillon A., React. Polym., **1991**, 16, 115.
22. Nakahara Y.; Kageyama H.; Nakahara F.; Doi Y., Daikoushi-kihou, **1989**, 40, 178.
23. Suzuki K.; Ottenbrite R.M.; Siddiqui J.A., Polym. Prepr., **1997**, 38, 335.
24. Ito Y.; Inaba M.; Chung D.; Imanishi Y., Macromolecules, **1992**, 25, 7313.
25. Tomalia D.A.; Naylor A.M.; Goddard W.A., Angew. Chem. Int. Ed. Eng., **1990**, 29, 138.
26. Newkome G.R.; Yao Z-Q.; Baker G.R.; Gupta V.K., Org. Chem., **1985**, 50, 2003.
27. Sanford E.M.; Frechet J.M.J.; Wooley K.I.; Hawker C.J., Polym. Prep., **1992**, 33(2), 654.
28. Yin R.; Ottenbrite R.M.; Siddiqui J.A., Polym. Prep., **1993**, 34(2), 506.
29. Nemeth S.; Ottenbrite R.M.; Siddiqui J.A., Polym. Prep., **1997**, 38, 365.
30. Yin R.; Ottenbrite R.M.; Siddiqui J.A., Polym. Prep., **1995**, 37(2), 265.
31. Iler R.K., The Chemistry of Silica; John Wiley & Sons: New York, 1979.
32. Yin R.; Ottenbrite R.M.; Siddiqui J.A., Polym. Prep., **1995**, 37(2), 262
33. Yin R.; Ottenbrite R.M.; Siddiqui J.A., Polym. Prep., **1998**, 39(1), 656.

APPLICATIONS

Chapter 14

Amine Functional Specialty Polymers as Adhesives in Multilayer Film Composite Structures

C. Chappell, Jr., E. L. Mason, and J. A. Siddiqui

Tech/R&D Laboratory, DuPont Polyester Films, P.O. Box 411, Hopewell, VA 23860

This chapter will provide an overview on the applications of amine functional polymers as coating materials in the preparation of multilayer composite film structures. In this study, amine functional polymers such as poly(ethylene imine), PEI, polyvinylamines, PVAm, polyvinylalcoholamines, PVAA, and Starburst® dendrimer, polyamidoamine (PAMAM) generation-2, were evaluated as interlayer adhesion promoting polymers between poly(ethylene terepthalate), and ionomeric polyethylene, Surlyn®. With the exception of Starburst dendrimer, all amino-functional polymers were found to be excellent adhesion promoters as tie coat adhesive layers between PET and Surlyn®. Using near infrared spectroscopy, NIRA, peel strength, and scanning electron microscopy data, we have shown that in the "PET-PEI-Surlyn®" composite structure, there is an amide bond between PET, and PEI interface, and metal-amine complexation at the PEI-Surlyn ® interface. Experiments also suggest ways for making strong interfaces in the "PET-PEI-Surlyn®" composite structures. The "PET-PEI-Surlyn®" interface integrity is very sensitive to the molecular weight, solution pH, chain conformation, side chain of the grafted poly(ethylene imine), and heat seal temperatures during the preparation of laminate composite structures.

In the flexible packing industry, composite film laminate structures of poly(ethylene terepthalate), (PET), with Surlyn®, and polyethylene are common. The laminate film structures combine the good properties of each individual substrate and maximize benefits from the standpoint of oxygen barrier, moisture barrier, flavor, heat seal, printing, gloss and aesthetics of a composite package. Without surface modifications, PET film doesn't adhere to Surlyn® or polyethylene. In order to promote adhesion between PET film and Surlyn ® or polyethylene (PE), amine functional polymers such as poly(ethylene imines), (PEI), grafted poly(ethylene imine) (g-PEI), polyvinylalcohol amines (PVAA), and polyvinylamine (PVAm) are used as a tie-coat adhesive layer in the preparation of composite film laminate structures (1,2,3,4). The amine functional polymers are applied as thin polymeric coatings on the PET film surface and molten polyethylene or molten Surlyn® is extruded as thin film on the amine functional modified PET. The "PET-PEI-Surlyn®" laminate can also be prepared by first coating the PET film surface with the amine functional polymer, followed by drying, and then finally heat sealing the coated PET film with Surlyn® or polyethylene film. In this study, we used Surlyn®, or polyethylene films in the preparation of the laminate composite structures.

An understanding of the structure, property relationship in the "PET-PEI-Surlyn®" interface is necessary in order to facilitate the design of laminate structure with desirable performance properties. Therefore, we wish to characterize the chemical nature of the poly(ethylene imine), and polyvinylamine in the tie coat layer between poly(ethylene terepthalate) and Surlyn®.

Our approach involves the selection of several types of commercially available amino functional polymers. In this study, we selected linear and branched poly(ethylene imine) functional polymers as a "tie coat layer" in the "PET-PEI-Surlyn®" composite laminate. It is our goal to correlate the peel strength in the "PET-PEI-Surlyn®" interface with the molecular characteristics of amino functional polymers.

Experimental

A) Materials: Several grades of poly(ethylene imines), trade name Lupasol P, and Lupasol SK were obtained from BASF Corporation; polyvinylalcoholamine (PVAA) and polyvinylamine (PVAm) were purchased from Air Products, Inc.; Starburst® dendrimer polyamidoamine was purchased from Dendritech Corporation. Poly(ethylene terepthalate) film , grade, Melinex® 800/48 and coextruded 850/48 films, and Surlyn®, grade 1702, were from DuPont Films.

B) Coating Method: 0.3%, 05%, 0.7% and 1.0% amine solutions were used for coating the PET films. The coating solutions were applied using Meyer bar, the coated surface was dried at180°C for one minute; the dry coating thickness range was between 0.01 micron to 0.05 micron.

C) Laminate Preparation: The PEI coated PET film was laminated with Surlyn® 1702 (200 gauge thick) at various temperatures between 290 to 450°F for one minute. The peel strength curves were obtained using Instron, and the plane of fracture was obtained using scanning electron microscopy(5). Near infrared spectra (NIRA) were obtained to study the surface amidation reaction of PET with PEI. The laminate structures are shown in figures 1 and 2. In figure 1, the PEI coating is on the semicrystalline surface of the PET film (Melinex® 800/48) whereas in figure 2, the PEI coating is on the amorphous side of the coextruded PET film (Melinex® 850/48). Both sides of the PET film (amorphous and crystalline) surface produce strong "PET-PEI-Surlyn®" interface.

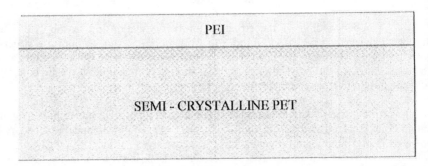

Figure 1: Poly (ethylene terepthalate) film, both sides are semi-crystalline semicrystalline and the second side is amorphous.

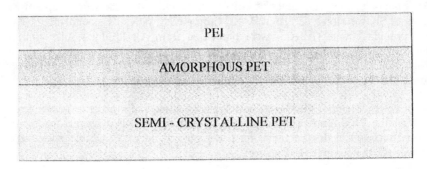

Figure 2: Poly (ethylene terepthalate) film, one side is semicrystalline and the second side is amorphous.

(D) Instrumentation

- Instron peel test was performed on an Instron, Instru-Met Model 1101.
- Heat seal laminates were prepared using a laminator, Sencorp System 12ASL
- Fourier transform infrared spectroscopy (FTIR) was performed on Nicolet Magna-IR/550.

Results and Discussion

In this section, we wish to describe: a) synthesis of PEI, PVAm, and PVAA using various monomers, b) effect of molecular weight on the bond strength of the PET-PEI-Surlyn interface, c) effect of coat weight on adhesion, d) effect of polymer composition on the bond strength at the PET-PEI-Surlyn interface and finally e) characterization of the PET-PEI-Surlyn® interface. Experiments suggest that polyvinylamines (PVAm) behave in a similar manner to PEI, whereas dendrimeric polyamidoamine has very poor tie-coat properties between PET and Surlyn®. In our discussions, we will refer to the composite laminated as "PET-PEI-Surlyn®" structure for the sake of simplicity.

A) Synthesis of PEI, PVAm, and PVAA

In figures 3, 4, and 5, commercial synthesis of poly(ethylene imine), polyvinylamine, and polyvinylalcohol amine are described (6,7).

B) Effect of Molecular weight on PET-PEI-Surlyn® Laminate Structures

In figure 6, the peel strength of the PET-PEI-Surlyn® interface as a function of molecular weight of PEI is shown. Clearly, data suggest that as the molecular weight of PEI increases, the peel strength also increases (8,9). For example, the low molecular weight PEI (M. wt. ~ 600 Daltons) gave a peel strength of 130 gms/inch^2 whereas the higher molecular PEI (M.wt ~ 50,000 Daltons) gave a peel strength of 580 gms/inch$^{2.}$ Interestingly, the polyvinylalcohol amine copolymer follows a trend similar to that of PEI. The hyper-branched, low molecular weight dendrimer behaves very much like a very low molecular weight PEI, and produced a weak interface between PET and Surlyn® films.

C) Effect of Coat Weight on the Peel Strength of PET-PEI-Surlyn® Interface

Several concentrations of PEI solutions (0.15%, 0.20%, 0.60% and 1.00%) were applied to the PET film surface. The coated film was dried at 180°C for 60 seconds. After coating and drying, the coated PET film was laminated with Surlyn® 1702, and the bond strength was measured using the Instron peel test method. Data suggest that as the coat weight increased; the bond strength at the PET-PEI-Surlyn® interface also increased. However, at higher PEI concentration levels (exceeding 1.0% PEI solution)

190

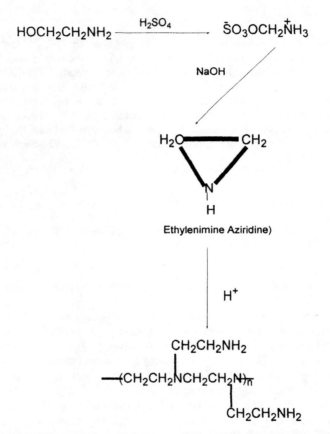

**Figure 3: Synthesis of Poly (ethylene imine), primary: secondary: tertiary
Nitrogen ratio: 1:1:0.75**

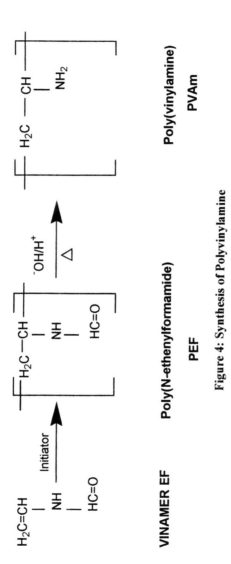

Figure 4: Synthesis of Polyvinylamine

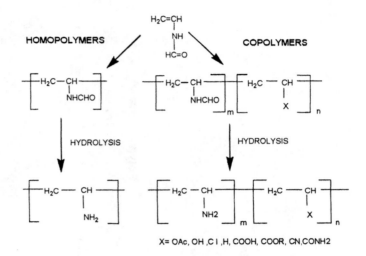

Figure 5: Synthesis of Polyvinylalcohol Amines (PVAA)

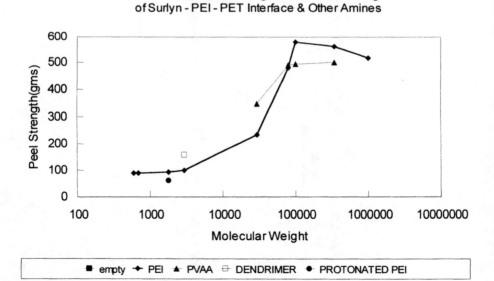

**Figure 6: Effect of Molecular Weight on the Peel Strength of PET-PEI-Surlyn®
Interface**

the bond strength begins to level out. Figure 7 shows the effect of coat weight on the peel strength of PET-PEI-Surlyn® interface.

D) Effect of PEI Composition on the Peel Strength of PET-PEI-Surlyn® Interface

In Figure 8, effect of PEI composition on the PET-PEI-Surlyn® interface is shown. Clearly, data suggest that protonated PEI (below pH = 7), (Figure 9) gives a very poor adhesive bond strength, whereas partially acylated PEI, (Figure 10), maintains good adhesive characteristics between PET and Surlyn® 1702.

E) Characterization of the Interfacial Bond between Poly(ethylene terepthalate) and Poly(ethylene imine):

In Figure 11, the near infrared (NIRA) data of the PEI coated PET is plotted. Clearly the presence of $-CONH_2$ group suggests an amidation reaction between the surface carboxylic group of PET and the amino functional groups of PEI during the curing process. In Figure 12, the effect of heat seal temperature on the peel strength is plotted. Interestingly, as the heat seal temperature increases, the peel strength also increases, suggesting an increase in the population density of amido groups. Additional experiments are in progress to quantify the increase in amido functional group concentration as a function of temperature. The peel strength data also suggest that as the lamination temperature increased above 400° F, the peel strength started to deteriorate This decrease in the peel strength is due to the degradation of PEI, resulting in the cohesive failure in the PEI region of the PET-PEI interface (Figure 10). A yellowing of the laminate was also observed above 400°F. Based on the peel strength, NIRA, and heat-seal data at different temperatures, we propose a chemical interface between PET and PEI via amidation reaction. The schematic of the interface is shown in Figure 13.

F) Characterization of the "Surlyn®-Poly(ethylene imine)" Interface.

Surlyn® is an ionically crosslinked thermoplastic derived from ethylene/methacrylic copoloymer; the ionically crosslinking metal ions are zinc and sodium. To investigate the nature of bonding between PEI and Surlyn®, several laminates were prepared using PEI under acidic and basic conditions. In Figure 8, data on the effect of pH on peel strength is plotted. Clearly, protonated PEI (pH below 5) has practically no bond strength, suggesting an "amine-metal" complexation reaction at the "Surlyn-PEI" interface. Since the protonated PEI prevents metal complexation, the interfacial bonding fails completely. In Figure 13, the schematic of the PEI-Surlyn® interface in the PET-PEI-Surlyn® interface is proposed.

Conclusions

High molecular weight amine functional polymers, such as PEI, PVAm and PVAA are excellent adhesion promoters and serve as a strong "tie-coat" in the preparation of multi-layer composite structures between poly(ethylene terephthate) and ionomeric Surlyn®. Using near infra-red spectroscopy, peel strength and heat seal data, we have

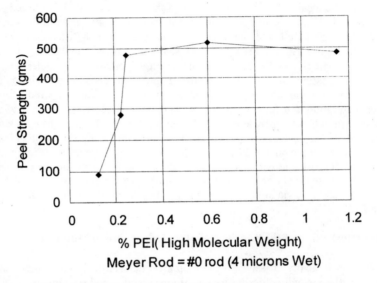

Figure 7: Effects of Coat Weight on PET-PEI-Surlyn® Interface,
MolecularWeight: 50,000 – 100,000 (Heat Seal Temperature = 300°F)

Peel Strength As A Function of PEI Composition

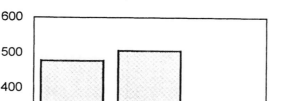

Figure 8: Effects of Composition on PET-PEI-Surlyn® Interface, Molecular Weight: 50,000 (Heat Seal Temperature = 380°F)

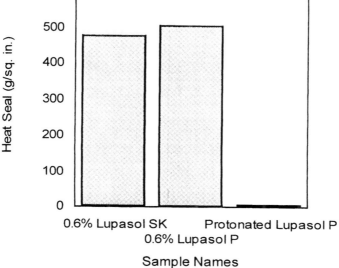

Figure 9: Effect of Composition on Peel Strength, Structure of Protonated Poly (ethylene imine), Poor Peel Strength.

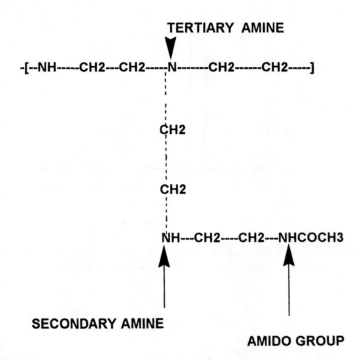

Figure 10: Acylated Poly (ethylene imine), Lupasol SK, Lupasol P has no -
COCH₃
Groups. See Peel Strength Data in Figure 8.

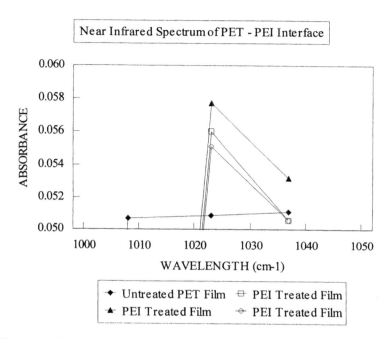

Figure 11: Near Infra-red Spectrum of the PET-PEI-Surlyn® Laminates

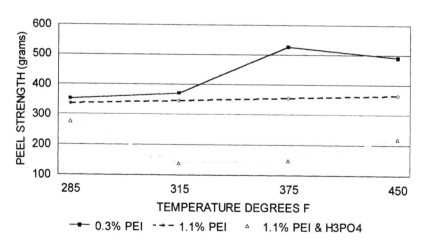

Figure 12: Peel Strength as a Function of Heat Seal Temperature

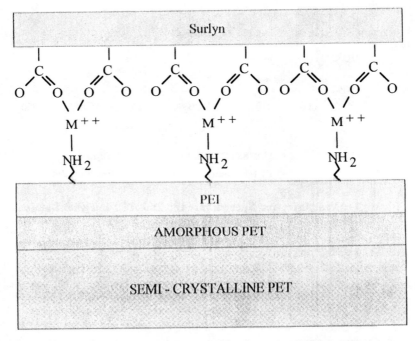

Figure 13: **Schematic Representation of PET-PEI-Surlyn® Interface**

shown that in the "PET-PEI-Surlyn®" composite structure, there is an amide bond formation between poly(ethylene terepthalate) and poly(ethylene imine) interface, whereas a strong "metal-amine" complexation occurs between poly(ethylene imine) and Surlyn® interface. The "metal-amine" complexation theory was substantiated by using protonated poly(ethylene imine) as a coating layer on the PET surface. Interestingly, use of protonated PEI on the PET surface has no peel strength with Surlyn®, resulting in a very poor PET-PEI-Surlyn® laminate structure. Additionally, the PEI coat thickness control at the PET-PEI-Surlyn® is very important. Initially, as the coat thickness increases, the peel strength increases, followed by leveling out of the heat seal strength, and finally the interface weakens due to the cohesive failure.

Acknowledgement

We would like to express our sincere thanks to the management of the E.I. DuPont de Nemours Company for giving us the permission to publish this work.

Literature Cited

1. Siddiqui, J.A., 1995, U.S. Patent #5, 453, 326

2. Siddiqui, J.A. Proceedings of the Seventh International Conference on Web Coating, Miami, Florida, November 10-12, 1993, Vol. 7, pp. 43-53

3. Siddiqui, J.A.; Mason, E.L., Chappell, Jr., C., *Polymer Preprint*, 1998, pp. 39, 111, 658-659

4. Siddiqui, J.A.; Mason, E.L.; Chappell, Jr., C.; *Polymer Preprint*, 1998 39(2), pp. 348-349

5. Kinloch, A. J.; Lau, C. C.; Williams, J. G., Int. J. Fract. 1994, 66, 45

6. DeRoo, A. M., *Handbook of Adhesives*, 2nd Edition, Editors Skiest, I., Van Nostrand, Rheinhold Company, 1977, pp. 592-596, Chap. 36

7. Winnik, A. M., Bystryak, M, S.: Liu, Z.; Siddiqui, J.A.; *Macromolecules* 1998, 31, 6855-6864

8. Lee, L, *Fundamentals of Adhesion*; Plenum Press, N.Y., 1991, Chaps. 1-4

9. Mittle, K. L., *Pure and Applied Chem.* 1980, Vol. 52, 1295

10. Belton, D. J.; Joshi, A, *Molecular Characterization of Composite Interface*, Marcel Dekker, Inc., N.Y. Edited by Ishida, H. and Kumar, G., 1988, p. 187

Chapter 15

Sodium 2-Acrylamido-2-methylpropanesulfonic Acid Polymers as Agents for Controlling Mist in Aqueous Fluids

Richard A. Denis, Sanjay Kalhan, and Steve R. Twining

The Lubrizol Corporation, 29400 Lakeland Boulevard, Wickliffe, OH 44092–2298

This chapter will focus upon the use of polymers as mist suppressants and on the development of tests which can measure mist suppression. Based upon these results, these polymers were added to metalworking fluids and tested in industrial machining operations for their effectiveness as mist suppression agents.

Aqueous fluids are extensively used in many commercial industries as agents for: process efficiency, friction modification, extending tool life, carrying away debris from a work surface, cooling surfaces, carrying additives, and providing other performance properties. During the course of these applications, these aqueous fluids commonly encounter shearing forces; such as high and low pressure spraying, high speed mixing, pumping and mechanical shear. The severity of these shearing forces causes part of the fluid to breakup into undesirable airborne particulates. The overall result is a loss of fluid, a reduction in processing, additional maintenance expenses, and creation of an unsafe work environment.

Reduction of mist could be accomplished via enclosures and/or upgrading existing ventilation systems at a considerable capital investment to these industries. Alternatively, mist suppression may be accomplished by use of chemical additives that are shear stable and capable of altering the size distribution of particles which are generated from the breakup of an aqueous fluid.

The use of additives in aqueous emulsions for mist suppression has been only recently been explored (1,2,3). In particular, high molecular weight polyethylene oxide (PEO) polymers have received considerable testing and have shown to have some degree of success in reducing mist. However, these polymers have a finite lifetime in the fluid before they shear degrade. Thus, this has led to the development of novel shear stable polymers derived from a highly hydrophilic sodium-2-acrylamido-2-methyl propane sulfonic acid (NaAMPS) monomer polymerized with hydrophobic monomers. (4)

Experimental

A) Materials:

Sodium-2-acrylamido-2-methyl propane sulfonic acid (NaAMPS) monomer was polymerized with a variety of hydrophobic alkylacrylate and alkylacrylamide monomers. The reactions were performed as solution polymerizations in methanol using the redox radical initiator of sodium persulfate and sodium metabisulfite.

B) Instrumentation:

The molecular weights were determined on a Waters 150 GPC using a set of Waters Ultrahydrogel 120, 250, 1000, and 2000 columns (7.8 x 300 mm), Mobile Phase was 80:20 water:acetonitrile with 0.10 M sodium nitrate, Flow Rate: 0.8 ml/min. (nominal), Column Temperature:35°C, Sample concentration: 1.6 - 2.0 mg/ml, Injection Volume: 200 μl, Detection: Differential Refractive Index, Sensitivity: -128, Calibration: narrow PEO and PEG standards, MW = 963,000 to 620 (Polymer Laboratories, Amherst, MA).

C) Spray Test:

The conditions for the Spray Analysis Test were set with a spraying time (t) = 10 seconds, an air pressure of 30 psig and a distance (x) at 38 cm. The solutions were pumped to the nozzle at a rate of 32 ml/min by means of a Sage model 355 syringe pump using a 50 cc syringe.

D) Grinder Test:

The Grinder Test machined a 1 inch square by 5 inch steel bar stock on an industrial grinder using a 0.5 inch by 8 inch grinding wheel at a removal rate of 1/10,000 of an inch per pass. The fluid pressure of the metalworking fluid was at 80 psi and utilized a 5 gal sump for the test fluid with a pump rate of 1.5 gallons per minute. The enclosed machining operation generated a mist which was measured in concentrations of mg/m^3 by a real time assessment monitor (DataRAM). The probes for aerosol measurement were located 14 inches and 24 inches from the grinding surface.

Results and Discussion

Development of a Spray Analysis Test for Measuring Mist Reduction

In metalworking operations, mist is the by-product from the atomization of a metalworking fluid which has been subjected to shear and extensional flow forces. A qualitative method to measure the reduction of aerosol mist was developed by incorporating the use of a spray nozzle as a source for continuous atomization of the fluid. A fluid that is sprayed through a nozzle at a set psi and at a given air to liquid ratio for a specified time will produce characteristic droplets which are analogous to mist.

By placing a screen at a distance (x) away from the nozzle, the aerosol droplets will impact on the screen and form a spray pattern (Figure 1). As the average aerosol droplet size decreases, the dispersion of the spray increases and the result is an increase in the size dimension of the spray pattern. A polymer which can alter the rheological properties of a solution such that it causes an increase in the average droplet size distribution produced by the nozzle will form a smaller spray pattern.

Figure 1. Spray Analysis Screen Test

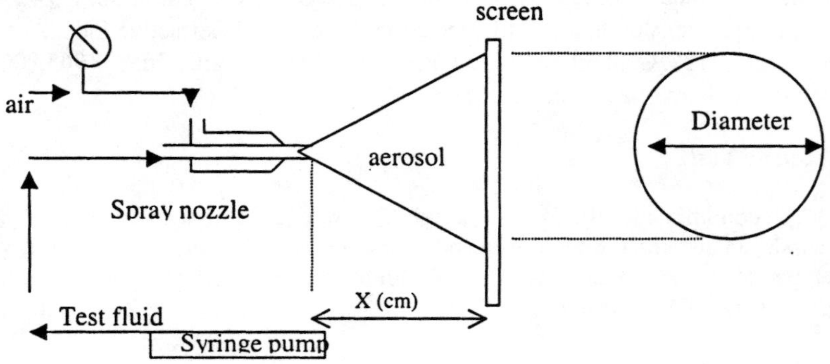

A qualitative method to determine the percentage of aerosol mist reduction can be easily obtained by measuring the diameter ($D_{additive}$) of the spray pattern from an "additized" aqueous fluid and comparing it to the diameter (D_{water}) of the spray pattern from the "water" baseline. The relative percentage of aerosol mist reduction can be calculated from these spray pattern as follows,

$$\Delta D_{(reduction)} = \frac{D \text{ water- } D \text{ additive}}{D \text{ water}} \times 100$$

Using the spray pattern analysis test, the amount of aerosol mist reduction, $\Delta D_{(reduction)}$ obtained for an aqueous solution containing 5000 ppm of a 1 million MW PEO polymer was calculated to be 20%, Figure 2. Similarly, the amount of aerosol mist reduction, $\Delta D_{(reduction)}$ obtained for an aqueous solution containing 5000 ppm of a 2 million MW PEO polymer was calculated to be 40%.

These polymeric solutions were sheared in a Waring blender at 25,000 rpm for 2 minutes in order to determine the effect that continuous shear has on the polymer's ability to suppress mist formation. The 1 million MW PEO solution was sprayed and the $\Delta D_{(reduction)}$ value was calculated to be < 6%, representing a 70% drop in mist suppression performance as compared to its non-sheared polymeric solution. The solution containing the 2 million MW PEO polymer had a $\Delta D_{(reduction)}$ of <3% which corresponded to a 92.5 % drop in mist suppression versus the performance of its non-sheared solution, as shown in Figure 2.

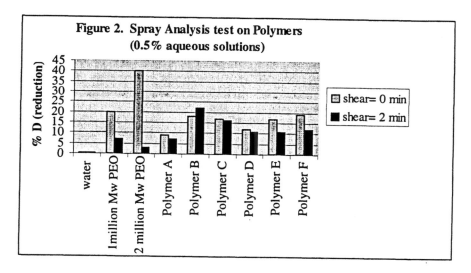

Figure 2. Spray Analysis test on Polymers (0.5% aqueous solutions)

Co-polymerization of NaAMPS with alkylacrylamides (Polymers A-D, Figure 2) and alkylacrylates (Polymers E-F, Figure 2) yielded polyelectrolytic polymers as antimist candidates. Aqueous solutions containing these copolymers were found to reduce to the spray pattern diameter, but to a lesser extent as compared to the 2 million MW weight polyethylene oxide. The series of Polymers A-F were tested in water at 5000 ppm and had $\Delta D_{(reduction)}$ values of 8-18% without shear. These polymer solutions were sheared in the Waring blender at 25,000 rpm for 2 minutes and then sprayed in the spray analysis test to determine their mist suppression performance. In contrast to PEO polymers, these polymers showed minimal loss in their ability to reduce mist. Polymers B and C had $\Delta D_{(reduction)}$ values of 18% and 17% before shear and $\Delta D_{(reduction)}$ values of 22% and 15% after shear. Polymer B showed a 22% improvement in mist suppression after shearing while polymer C had a nominal 12% loss in mist suppression performance as compared to their non-sheared solutions. Similarly, Polymers A and D experienced 20% and 15% loss in mist suppression performance as compared to their non-sheared solutions. The alkylacrylate polymers (E-F) were not as sheared stable, experiencing respectively 46% and 35% drop in mist suppression performance vs non-sheared solutions, as compared to their alkylacrylamide counterparts but they were significantly more shear stable than PEO polymers. The spray analysis test results were within 10% reproducibility.

Effects of Blender Shear on the Molecular Weight of the Polymer

The ability of PEO polymers to function as mist suppressants is directly related to the average molecular weight of the polymer. Both the 1 million and 2 million MW PEO polymers, as shown in Figure 2, had a significant loss in their ability to suppress mist after being sheared in the Waring blender. In contrast, the polyelectrolyte copolymers maintained a significant portion of their mist suppression capabilities even after experiencing severe mechanical shear. Molecular weight measurements on polymer samples before shearing and after shearing were determined by use of gel permeation chromatography.

Aqueous solutions containing 2%w of a 1 million MW PEO polymer, 600,000 MW PEO, 300,000 MW PEO, Polymer B and Polymer D were each sheared in a Waring Blender at 25,000 rpm. The 300,000 MW and 600,000 Mw PEO samples were included in the test as a comparison to the polyelectrolyte polymers that had molecular weights within this range. Samples of each polymer solution were collected after 2,5,15, and 30 minutes of shear. The molecular weights of the samples were determined by GPC analysis and plotted versus their shearing time. Figure 3 shows that after 2 minutes of blender shear the 1,000,000 MW PEO polymer had decreased to a 53,000 MW polymer (a 90% loss in MW) and both the 300,000 MW and 600,000 MW PEO polymers had lost virtually their total molecular weight.

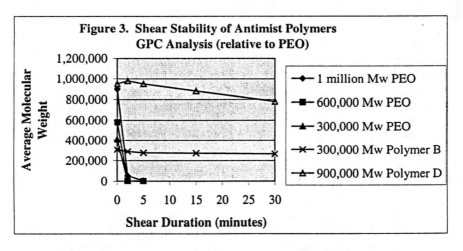

Figure 3. Shear Stability of Antimist Polymers GPC Analysis (relative to PEO)

The GPC molecular weight of Polymer B before shearing was determined to be 309,000 MW (relative to the PEO standard). Shearing this sample for 2 minutes resulted in a decrease in its molecular weight to 277,000 MW (<10% loss), after 5 minutes it was reduced to 265,000 MW (15% loss) and after 30 minutes of shearing it was at 200,000 MW (33% loss). The GPC molecular weight of Polymer D was determined to be 955,000 MW and after 30 minutes of shearing it had dropped to 780,000 MW (18% loss).

These results clearly show that PEO polymers are shear unstable and do not significantly suppress mist under conditions where a shear force is applied; whereas, the polyelectrolyte polymers prepared from NaAMPS are shear stable and maintain both their mist suppression performance and molecular weight under extreme shearing force conditions.

Correlation Between the Spray Analysis Test and Grinder Antimist Performance

A grinder test was developed to correlate the mist reduction observed in the spray pattern analysis test with an actual metalworking operation. Both a semisynthetic which contains 10% emulsified oil and a synthetic metalworking fluid which contains no mineral oil, were used during the machining of steel bar stocks.

Semisynthetic Metalworking Fluids

The amount of aerosol mist was established for the baseline semisynthetic emulsion and was assigned the value of 0% mist reduction. Additizing this fluid with 1000 ppm of the 1 million MW PEO resulted initially in a 70% reduction in mist (Figure 4); however, after 0.5 hr of machining the mist level increased by 10% over the baseline fluid. Additional machining for 1.0 hr resulted in further degradation of the polymer and an increase in mist of 20% over the baseline. The spray analysis test predicted <6% mist reduction.

Similar mist increases where observed when the semisynthetic fluid was treated with either 1000 ppm of 600,000 MW and 300,000 MW PEO polymers. Both the 600,000 MW PEO and 300,000 MW PEO polymers initially provided 40% mist reduction, but after 1.5 hr of machining the mist measured for both was 10% higher than the baseline (Figure 4).

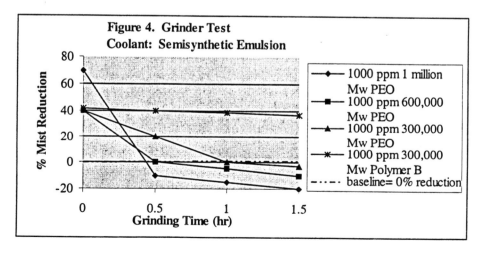

Figure 4. Grinder Test
Coolant: Semisynthetic Emulsion

The semisynthetic metalworking fluid was additized with 1000 ppm of Polymer B. The additized solution provided 45% mist reduction upon the initial addition of the polymer (Figure 4) and after 1.5 hr of machining there was >35% mist reduction. This result correlates with the spray analysis test which predicted 22% mist reduction

Synthetic Metalworking Fluid

The amount of aerosol mist was established for the baseline synthetic metalworking fluid and was assigned the value of 0% mist reduction. The synthetic metalworking fluid was additized with 1000 ppm of a 1 million MW PEO polymer. In this fluid, this PEO polymer caused a 100% increase in mist within minutes upon addition to the machining operation (Figure 5). Treatment of this metalworking fluid with 1000 ppm of the 300,000 MW PEO polymer caused a 60% increase in the amount of mist.

The synthetic metalworking fluid was additized with 1000 ppm of Polymer D and which provided an immediate 60% reduction in mist and this was maintained throughout the 1.5 hr of machining (Figure 5).

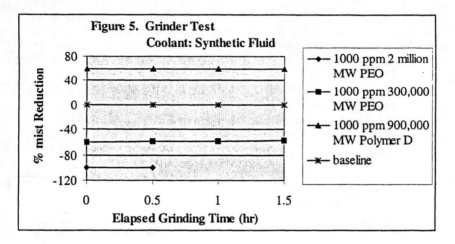

Summary of Polyethylene Oxide Polymers as Mist Suppressants

The differences in mist suppression performance for the polymer types tested may be attributed to either the solubility variations of the polymers in the two types of metalworking fluids and/or to interactions of the polymer with components in the metalworking fluid. PEO polymers had initial mist suppression performance in the 10% semisynthetic fluid even though there was a loss in performance after short time period of machining. However, in synthetic fluids these polymers generated more mist than the baseline. The solubility of the PEO polymers in these fluids was not determined; therefore, it can not be ascertained whether the polymers were in solution or if polymeric interactions or lack of interactions with the fluid inhibited full chain extension for these polymers.

Summary of Polyelectrolyte Polymers as Mist Suppressants

The performance of the polyelectrolyte polymers were not negatively affected by either type of fluid. Compositionally Polymer D is higher in molecular weight and more hydrophilic than Polymer B. The mist suppression capabilities of Polymer D may be partly attributed to elongation of the polymer during spraying; however, the hydrophilic Polymer D may be more soluble in the synthetic fluid and form stronger polymer to polymer associations. Controlling particle size of the fluid during a shearing force may be the result of these associations breaking rather than the polymer.

Likewise, the hydrophobic portion of Polymer B could interact with the oil that is in the semisynthetic metalworking emulsion to form an aggregate. The individual aggregates could form large associations and remain stable within the emulsion. The mechanism for shear stability for both Polymer D and Polymer B may be the result of these aggregates being broken during a shearing force and then during a relaxation state the aggregates recombine. Therefore, instead of fluid particles being subjected to a shearing force and forming mist particles, it is these large aggregates that prevent the break up of the fluid.

Industrial Field Trials

Field trials for determining the mist suppressing effectiveness of Polymer B and the 2 million MW PEO polymer were performed using a high pressure CNC machine at a major automotive plant. The operation of the CNC machined was designed to test the stability of antimist polymers under high pressure shearing. A high pressure (500 psi) stream of a metalworking fluid containing an antimist candidate was directed against a tool holder but without machining. The fluid was recirculated at 50 gallons per minute with a sump capacity of 120 gallons. Mist reductions were calculated from the mist that was measured by the DataRAM for the non-polymer containing baseline versus the mist that was measured for the baseline fluid that had been treated with the polymer.

Semisynthetic Metalworking Fluids

A 10% semisynthetic metalworking fluid was additized with 200 ppm of a 2 million MW PEO polymer and Figure 6 depicts the percentage of mist reduction versus high pressure spraying time. Initially, 40% mist suppression was measured but within 0.5 hr this dropped to a 60% increase in measured mist. An additional 300 ppm of the 2 million MW PEO polymer increased the level of mist to 70-80% over the baseline.

In comparison, Polymer B was tested in the 10% semisynthetic fluid at 500 ppm, Figure 6. Polymer B provided 25% mist suppression which held constant for 0.60 hr.

Afterwards, an additional 500 ppm of Polymer B improved the mist suppression level to 40% and this remained constant throughout the 9 hr of testing.

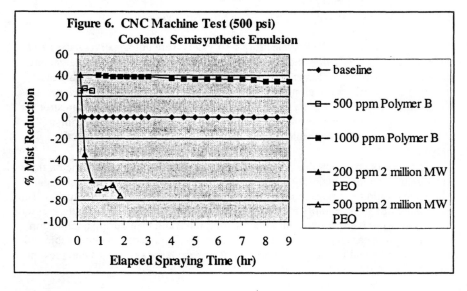

Figure 6. CNC Machine Test (500 psi)
Coolant: Semisynthetic Emulsion

Soluble oil Metalworking Fluid

Another CNC tests was conducted with a soluble oil emulsion using the same high pressure spraying conditions. The baseline for the soluble oil was established; afterwards, the soluble oil was additized with 200 ppm of the 2 million MW PEO polymer. The initial mist reduction measured was 85% less mist than the baseline but after 7 hr of shear degradation through spraying, this level had fallen to 10% mist reduction, Figure 7.

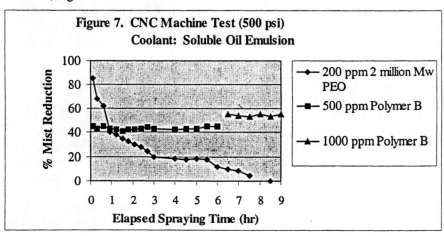

Figure 7. CNC Machine Test (500 psi)
Coolant: Soluble Oil Emulsion

Polymer B was also tested in the soluble oil emulsion and at a 500 ppm treat rate in the fluid it provided 45% mist reduction as shown in Figure 7. The high pressure spraying continued for 6.0 hr with no loss in mist suppression performance. The level of Polymer B was increased to 1000 ppm and the amount of mist suppression improved to 55% and this remained constant until the completion of the test.

Conclusion

Spraying and atomization of fluids involves extensional or stretching flow fields; therefore, increasing the extensional viscosity of the fluid is expected to increase the breakup length and thus the average droplet size of the spray. In systems which have minimal shear, PEO polymers function well as mist suppressants because they have the extensional viscosity of the fluid (5). The extensional viscosity of PEO solutions decrease when the solutions are subjected to shear; therefore, as seen in the above tests, a high molecular weight PEO polymer had initially provided mist suppression but as the polymer was sheared and its molecular weight decreased its ability to suppress mist was lost.

The mechanism by which the polyelectrolyte polymers functioned as mist suppressants differs from that of the PEO polymers. Polymer D which had an average molecular weight 900,000 MW has sufficient size to reduce mist by a elongation of the polymer in solution. Although, this polymer is able to suppress mist by this mechanism, it does not shear degrade like the corresponding 1,000,000 MW PEO polymer which was found to be shear unstable. This would indicate that Polymer D may not be operating solely by increasing the extensional viscosity of the solution, but instead it may function as a mist suppressant by means of an association mechanism. This association mechanism is further supported by the fact that Polymer B had an average molecular weight of 300,000 MW which is sufficiently low that it would have limited effect upon the extensional viscosity of the fluid and would not be able to suppress mist by any significant amounts. However, Polymer B was capable of suppressing mist despite its relatively low molecular weight; therefore, the majority of mist suppression can be attributed to the polyelectrolytic nature of these polymers. This polyelectrolytic nature of these polymers would favor the formation of aggregates between the polymer and the components within the metalworking fluid. These aggregates would then associate and under a shear force condition, these associations would break apart rather than the polymer. Therefore particle size control (mist suppression) for these polyelectrolyte polymers may be a combination of both an association mechanism and a visco-elastic mechanism.

Acknowledgments

The authors would like to express our thanks to the Lubrizol Corporation and the Ford Motor Company for providing us permission to publish the results of this work.

210

Literature

1. Gulari,E; Manke, C.W, Smoleski, J.M., Kalhan, S.N: Effect of Polymer Additives on the Atomization of Emulsions, Paper No. 97g, American Institute of Chemical Engineers Annual Meeting, San Francisco, CA (1994)

2. Kalhan, S. N.: Effect of Polymer Surfactant Interactions on the Atomizations of Emulsions, Ph.D. Dissertation, Wayne State University (1995)

3. Smolinski, J.M.: Polymeric Additives for Suppression of Machining Fluid Mists, Society of Manufacturing Engineers Metalworking Fluids Seminar, Troy MI (1996)

4. Kalhan, S.N.; Mann,J.;T; Denis,R.A.: Sulfonate Containing Copolymers as Mist Suppressants in Water-Based Metalworking Fluids. European Publication Reference 08116772A2, (1997)

5. Soyles, D.A.; Dinga, G.P.; Glass, J.E.: Dynamic Uniaxial Extensional Viscosity. Response in Spray Applications, Polymers as Rheology Modifiers, ACS # 462 (1991)

Chapter 16

New Saccharide-Derived Monomers and Their Use in Water-Treatment Polymers

J. S. Thomaides[1], P. M. Petersen[1], R. W. R. Humphreys[1], A. M. Carrier[2], A. Austin[2], and R. Farwaha[3]

[1]National Starch and Chemical Company, 10 Finderne Avenue, Bridgewater, NJ 08807
[2]ALCO Chemical, 909 Mueller Drive, Chattanooga, TN 37406
[3]Nacan Products Limited, Brampton, Ontario L6T 4W7, Canada

Carbohydrates ranging from monosaccharides to large oligosaccharides were functionalized with a single polymerizable α-methyl styryl group using reductive amination as the key synthetic step. These new monomers were then copolymerized with acrylic acid to give water treatment polymers with pendent saccharide functionality. The thus modified water treatment polymers could be detected down to 1 part per million using a classical quantitative analytical test for saccharides.

Saccharides have a number of attributes that make them very attractive as raw materials for the synthesis of polymers. The confluence in saccharides of different functionalities such as multiple hydroxyl groups and latent reactivity, which is difficult to realize in wholly synthetic materials, is of particular interest to us. The preparation of monomers derived from saccharides and the subsequent polymerization of these materials is one approach that has been extensively pursued as a means to introduce saccharide groups into synthetic polymers (1-9). With a few exceptions (2), most of this previously reported work has involved attaching a polymerizable moiety onto a mono- or disaccharide. The practical synthesis of a new family of monomers derived from carbohydrates ranging from monosaccharides to large oligosaccharides and the use of these monomers to produce a detectable water treatment polymer are described in this paper.

Experimental

Materials. All reagents were used without further purification. Corn syrup solids (DE 24) and maltodextrin (DE 10) were obtained from American Maize-Products

Company. Oligosaccharide **1g** is a product of National Starch and Chemical Company. United Catalysts G-49B was provided by United Catalysts Inc. All other reagents were commercially available from multiple sources. Ion exchange SPE tubes (LC-SAX, quaternary amine, chloride counter ion) were obtained from Supelco.

Instrumentation. Proton and carbon-13 NMR were obtained using a Bruker AM-300 or AC-250 instrument. UV/VIS spectra were recorded on a Perkin Elmer Lambda 5 Spectrophotometer using a 10 cm (cylindrical) pathlength cuvette.

Molecular weight determination was done by gel permeation chromatography (GPC) using a Waters liquid chromatograph system equipped with a refractive index (RI) detector. The conditions employed were 0.05M $NaH_2PO_4 \cdot H_2O$ and 0.05M Na_2HPO_4 (in H_2O) mobile phase; 1000 p.s.i. pressure; a flow rate of 1.0 ml/min; a PROGEL TSK PW column series consisting of Guard PWXL, 4000 PWXL, 3000 PWXL, and 2500 PWXL columns; and a temperature of 35°C. Poly(acrylic acid) standards were used for calibration.

Exemplary Reductive Amination - Synthesis of N-Methyl-D-Lactamine. α-D-Lactose monohydrate (100 g, 0.28 mole) was dissolved in deionized water (150 ml) with the aid of stirring and heating. The solution was cooled to room temperature then added over two hours to a solution of 40% (w/w) methylamine (43 g, 0.55 mole) in deionized water (50 ml) while stirring under an atmosphere of nitrogen and holding at 0-10°C. The resultant mixture was stirred for a further one hour, and then it was poured into a pressure reactor along with United Catalysts G-49B (10 g). The pressure vessel was then heated at 55°C under an atmosphere of 700 p.s.i. hydrogen for 24 hours. After this time, the reaction vessel was depressurized, and the catalyst was removed by filtration. A small sample of the filtrate was evaporated to dryness under dynamic vacuum in order to obtain a sample for characterization. The filtrate was evaporated to near dryness to ensure the removal of excess methylamine. Amine concentration was determined by titration against 0.1 N HCl. Proton NMR (in D_2O): 4.3 ppm (d, 1H), 3.9 (m, 1 H), 3.7-3.3 (m, 11 H), 2.6 (m, 1H), 2.4 (m, 1H), 2.2 (s, 3H). Carbon-13 NMR (in D_2O): 105 ppm, 81, 77, 75, 73.5, 73.2, 73.1, 72, 71, 64, 63, 55, 37.

Exemplary Monomer Synthesis. To a solution of N-methyl-D-lactamine (95 g, 0.27 mole) in deionized water (350 ml) was added 3-isopropenyl-α,α-dimethylbenzyl isocyanate (54 g, 0.27 mole), and the resultant two phase mixture was vigorously stirred for ten hours. The course of the reaction was monitored by observation of the disappearance of the isocyanate peak at ~2250 cm^{-1} in the infrared spectrum of the reaction mixture. A small amount of an off-white precipitate formed and was removed by filtration. Unreacted aminosaccharide (<5%), as determined by titration of the reaction mixture versus dilute hydrochloric acid, was removed by treatment with Amberlite IR-120 (plus) ion exchange resin. The filtered, amine-free solution was freeze-dried to yield a white solid (143 g, 96% of theory). Proton NMR (in D_2O): 7.5 ppm (s, 1H), 7.3 (m, 3H), 5.4 (s, 1H), 5.1 (s, 1H), 4.5 (d, 1H), 4.0-3.2 (m, 12 H), 2.9 (s, 3H), 2.1 (s, 3H), 1.5 (s, 6H). Carbon-13 NMR (in D_2O): 157 ppm, 146,

142, 139, 126, 122, 121, 120, 110, 101, 79, 73, 71, 70, 69, 67.9, 67.8, 66.5, 60, 59, 53, 49, 33, 28, 19.

Exemplary Copolymerization. To a reaction vessel was charged deionized water (135 g), 5 ppm of an iron salt, and isopropanol (52.2 g). Into three separate feed containers were charged: (1) acrylic acid (102.6 g, 1.5 mole); (2) monomer **5c** (27.9 g, 0.05 mole), isopropanol (20 g) and deionized water (30 g); and (3) sodium persulfate (10.44 g 0.046 mole) and deionized water (50 g). The three feed containers were attached to the reaction vessel which was heated to ~85°C. The sodium persulfate solution was slowly added at such a rate as to give uniform addition over 3.25 hours. After 0.25 hours, the acrylic acid and monomer **5c** were added at a rate such as to give uniform addition over three hours. The isopropanol was distilled from the reaction mixture after all solutions had been added. The reaction was then neutralized with a 50% sodium hydroxide solution (75 g). The resulting polymer had a weight average molecular weight of 6500 and a polydispersity of 2.06. Samples of aqueous polymer solution were freeze-dried and analyzed by NMR.

Phenol-Sulfuric Acid Colorimetric Procedure. A 4 ml sample of saccharide solution was pipetted into a test-tube, and 0.2 ml of 80% phenol solution was added. To this mixture was added 10 ml of concentrated sulfuric acid rapidly. The stream of the acid was directed against the liquid surface rather than against the side of the test-tube in order to obtain good mixing (caution: the solution and test-tube become very hot during the addition of acid). After the solution stood for 30 minutes, the absorbance at 485-490 nm (preferably 487 nm) was measured.

Solid Phase Extraction (SPE) Method for Detection of Saccharide-Tagged Polymers. An SPE tube containing ~0.5 g of LC-SAX (strong anion exchange) packing material was conditioned with methanol (2 ml) and then water at pH 7-10 (5 ml). A 100 ml sample of the saccharide-tagged polymer solution at pH 7-10 was then passed through the packing material by syringe. The packing material was then washed with water at pH 7-10 (2 X 5 ml). The polymer was eluted from the packing material by passing dilute HCl (pH ~1; 2 X 2 ml) through the column. The 4 ml sample thus collected was developed with phenol-sulfuric acid as above.

Results and Discussion

Monomer Synthesis. Saccharides can readily be modified by reactions centered on their hydroxyl groups. The challenge in introducing a single polymerizable functionality to a saccharide arises from the near equivalence, with respect to reactivity, of the multiple saccharide hydroxyls. One strategy adopted by some groups to differentiate the saccharide hydroxyls is the protection of all but one hydroxyl, which is then used as a point of attachment for a polymerizable moiety (*3,4*). Subsequent deprotection is then necessary in order to regain the saccharide character of the starting material. In another approach, enzymes are used to selectively introduce polymerizable functionality at a single site (*5,6*). An alternative approach is to introduce to the saccharide by reductive amination a single amine functionality, which serves as a reactive handle for further modification (*7-9*). We chose to follow the last approach because it appeared to be the most general and best suited for the preparation of monomers derived from oligosaccharides.

Our two step synthesis of saccharide derivatives with a single, polymerizable α-methyl styryl group is shown in Figure 1. In the first step, a saccharide with a single reducing end group (**1**) is reductively aminated with methylamine (**2**) to give the corresponding monoamino saccharide (**3**). The process involves hydrogenation of a mixture of saccharide and amine in water over a supported nickel catalyst in water. Reductive amination of sugars is well known; catalytic hydrogenation (*7-11*) or chemical reduction (*12*) have been used to effect this transformation. We found that catalytic hydrogenation, which appears to have been previously reported only for mono- and disaccharides, could readily be applied to maltodextrins with as many as about 50 anhydroglucose repeat units. Thus, amino saccharides **3a-g** were readily obtained from saccharides **1a-g** in a fully aqueous reaction without the generation of any by-products that would be associated with a chemical reduction. Proof of structure was obtained by proton and carbon-13 NMR and titration analysis. It should be noted that corn syrup solids (DE 24), **1e**, maltodextrin (DE 10), **1f**, and oligosaccharide **1g**, which are all obtained by controlled enzyme or acid catalyzed degradation of corn starch, are polydisperse.

Treatment of amino saccharides **3a-g** with 3-isopropenyl-α,α-dimethylbenzyl isocyanate, compound **4**, in water with vigorous agitation produced carbohydrates **5a-g**, which contain a polymerizable α-methyl styryl moiety. The reaction was highly selective; no evidence for the condensation of **4** with a saccharide hydroxyl could be found. The isolated yield of highly pure monomer over the two steps ranged from 85-95%. The condensation reaction is interesting in that isocyanate **4** is not water soluble; presumably, reaction of it with the water soluble amino saccharide occurs at the interface between the aqueous solution and the insoluble droplets of isocyanate.

Polymerization Studies. In order to evaluate the polymerization aptitude of our saccharide derived monomers, we attempted the copolymerization of lactose derived monomer **5c** with acrylic acid. Saccharide functionalized low molecular weight poly(acrylic acid) was of particular interest to us (see below). In a typical run, a comonomer feed consisting of 21.3 wt.% (3.4 mole%) lactose derived monomer **5c** and acrylic acid was polymerized in aqueous isopropanol using sodium persulfate as the initiator. Conversion of both monomers as determined by proton NMR was essentially quantitative. In Figure 2, matching portions of the proton NMR spectra of monomer **5c** and a copolymer of **5c** with acrylic acid are reproduced. In the spectrum of the monomer, Figure 2a, resonances at 5.4 and 5.1 ppm that are characteristic of the vinylic protons of **5c** are clearly evident as are resonances at 7.5 and 7.3 ppm, which are characteristic of the aromatic ring protons of monomer **5c**. In the spectrum of the copolymer, Figure 2b, the vinylic resonances of **5c** (and acrylic acid) are essentially absent while a single broad resonance in the aromatic region is present. These observations strongly suggest that both comonomers were converted to a high

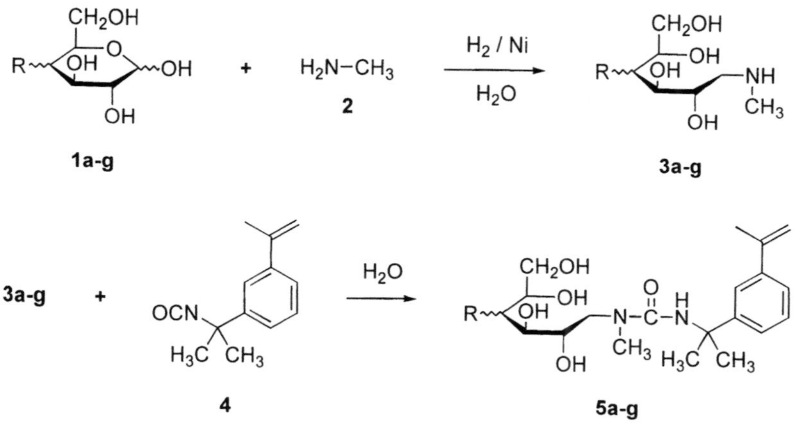

a: R = α-OH
b: R = β-OH
c: R = α-[β-D-galactopyranosyl]
d: R = α-[α-D-glucopyranosyl]
e: R = α-[α-D-glucopyranosyl-(1→4)-]$_{\sim4}$
f: R = α-[α-D-glucopyranosyl-(1→4)-]$_{\sim10}$
g: R = α-[α-D-glucopyranosyl-(1→4)-]$_{\sim50}$

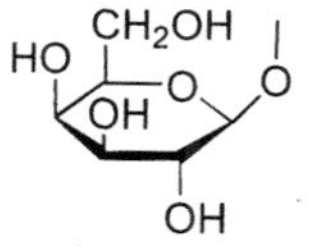

β-D-galactopyranosyl

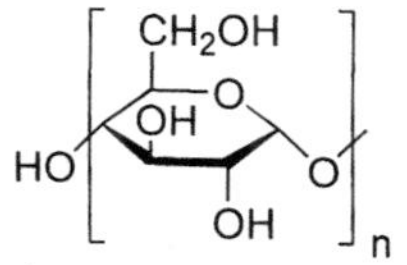

[α-D-glucopyranosyl-(1→4)-]$_n$-

Figure 1. Two-step synthesis of saccharide monomers.

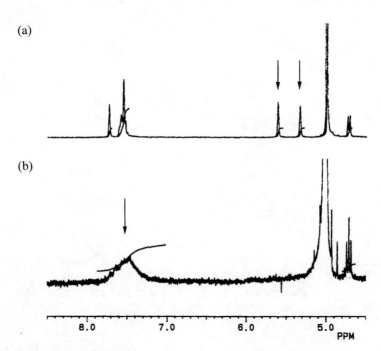

Figure 2. Proton NMR spectra of monomer **5c** (a) and copolymer made with monomer **5c** and acrylic acid (b).

degree in the copolymerization. Similar results were obtained with saccharide monomers **5e** and **5f**. The feed of these monomers were lower (typically on the order of 1-2 mole%), so NMR could only provide a qualitative measure of the incorporation of the saccharide monomer into the copolymers in these cases. It was, however, possible to get a more quantitative measure of the incorporation of monomers **5e** and **5f** into the acrylic acid copolymers using a solid phase extraction technique (see below).

Labeled Water Treatment Polymer. Low molecular weight poly(acrylic acid) is used as a dispersant for mineral scale in water treatment applications. The water treatment polymer is added to the aqueous system in a predetermined concentration that is effective to inhibit the formation and deposition of mineral scale. The concentration of the polymer must be monitored over time in order that the amount of polymer present in the system can be maintained at the predetermined concentration.

Low molecular weight poly(acrylic acid) is difficult if not impossible to detect directly in aqueous solution at the concentrations at which it is employed in water treatment applications. There are, however, many classic colorimetric techniques for the quantitative detection of saccharides down to the ppm level (*13*). If a poly(acrylic acid) water treatment polymer is prepared with pendent saccharide functionality, detecting the polymer reduces to a problem of detecting the saccharide functionality. With monomers of the type **5** in hand, we were able to explore this strategy for the preparation of a detectable water treatment polymer.

Using the phenol-sulfuric acid test for carbohydrates (*14*) we were able to detect monomers **5c-f** at concentrations down to 1 ppm. In this test for saccharides with glycosidic linkages or reducing end groups, the saccharide is mixed with phenol and concentrated sulfuric acid. The presence of saccharide is indicated by the formation of an orange color, and the amount of saccharide can be quantified by measuring the peak absorbance at 485-490 nm and comparing the experimental value to a suitable standard. As is the case for many saccharides, the response of the absorbance at 485-490 nm versus saccharide monomer concentration from 1 to 100 ppm was linear in accordance with the Beer-Lambert Law, which makes the phenol-sulfuric acid test a useful tool for detecting and quantifying saccharide monomers **5c-f** in water at these levels. Consistent, with expectations, monomers **5a** and **5b**, which have no glycosidic linkages, could not be detected by the phenol-sulfuric acid test.

A copolymer consisting of monomer **5f** (2 mole%) and acrylic acid could be readily detected down to a concentration of 10 ppm using the phenol-sulfuric acid test. In a control experiment, unmodified poly(acrylic acid) gave a negative phenol-sulfuric acid test. The loss of sensitivity of the method on going from the monomer to the copolymer is the result of "dilution" of the saccharide moiety with acrylic acid in the copolymer.

Although detection down to levels of 10 ppm is of some value, it would be more useful to have a method capable of detecting the polymer down to 1 ppm. This readily could be done by concentrating the solution of the tagged polymer prior to subjecting it to the phenol-sulfuric acid test. A 25-fold concentration was accomplished by passing 100 ml of dilute polymer solution (1-10 ppm polymer) through a column of strong anion exchange resin conditioned at pH 7-10. The copolymer is retained on the column because of its anionic nature (at pH 7-10). The copolymer was then flushed from the column using 4 ml of dilute HCl (pH ~1). At low pH, the copolymer is neutral, so the resin no longer retains it. In Figure 3, absorbance of the concentrated polymer solution after development with phenol-sulfuric acid is plotted versus the concentration of the original polymer solution. Clearly, the copolymer can be detected down to 1 ppm using this methodology. It is significant that the response of absorbance versus concentration is linear in the copolymer system in the concentration range studied as was the case with the corresponding saccharide monomer. In addition to making the determination of polymer concentration possible, the linear response suggests that the polymer is quantitatively trapped on the ion exchange resin at high pH and then quantitatively released at low pH.

The fact that the neutral saccharide residue is trapped on the column and subsequently can be eluted as though it contains carboxylate functionality further supports the thesis that the saccharide monomer is incorporated into the poly(acrylic acid). In a control experiment, glucose was not retained on the ion exchange column in the presence of the copolymer while the copolymer was. These observations were used to advantage in the determination of saccharide monomer incorporation into the acrylic acid copolymer. A solution of polymer made from a feed containing 2 mole% monomer **5f** that was not otherwise purified was subjected to the solid phase extraction technique. Forty-six percent of the saccharide present in the copolymer solution prior to solid phase extraction was not retained on the solid phase extraction medium; 54% of the saccharide was retained and therefore bound to the copolymer. This indicates that a feed of 2 mole% of this saccharide monomer leads to the incorporation of 1.1 mole% monomer in the copolymer. Similar levels of monomer incorporation were observed in other copolymerizations of monomer **5f** and with monomer **5e** (*15*).

The effectiveness (and efficiency) of two saccharide labeled water treatment polymers as calcium scale inhibitors were then evaluated against low molecular weight poly(acrylic acid), which served as a control. The results of this study are summarized in Table I. The scale inhibition of the labeled polymers was comparable to that of the control at 5 ppm. This indicates that the saccharide tag has no adverse effect on the water treatment performance of the polymer other than, perhaps, to dilute the amount of acrylic acid residues present per unit of polymer present (note the drop in the performance of the experimental polymers relative to that of the control polymer at a concentration of 3 ppm).

Conclusions

A very practical and broadly applicable two-step synthesis of a new family of mono-functional saccharide monomers was developed (*16*). In this synthesis, all reactions are done in water, no protecting groups are employed, and no by-products are formed. These saccharide-derived monomers were found to be useful as "tags" for water treatment polymers (*17*).

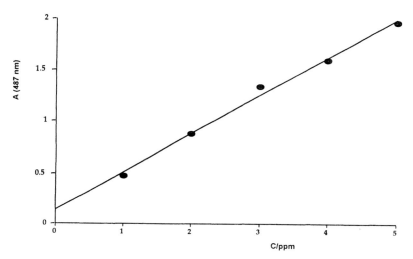

Figure 3. Detection of saccharide modified poly(acrylic acid).

Table I. Evaluation of saccharide labeled poly(acrylic acid)

Polymer	Treatment level (mg/L)	Average % Inhibition[a]
Poly(acrylic acid)	1	39.2
	3	98.6
	5	98.6
	10	95.6
Acrylic acid (98 mole%) /	1	8.9
5f (2 mole%) copolymer[b]	3	64.3
	5	97.2
	10	95.6
Acrylic acid (97 mole%) /	1	9.9
5e (3 mole%) copolymer[b]	3	56.0
	5	93.3
	10	93.9

a. Reported as % stabilization of calcium carbonate. b. Monomer feed ratio.

Acknowledgments

The authors would like to thank Dr. Richard E. Zadjura and Dr. Anthony L. Romanelli of National Starch and Chemical Company for assistance with NMR spectroscopy, Susan B. Feeser of ALCO Chemical for GPC analysis, and National Starch and Chemical Company for its support and interest.

Literature Cited

1. Wulff, G.; Schmid, J.; Venhoff, T. *Macromol. Chem. Phys.* **1996**, *197*, 259-274.
2. Kobayashi, K.; Kamiya, S.; Enomoto, N. *Macromolecules* **1996**, *29*, 8670-8676.
3. Jhurry, D.; Deffieux, A.; Fontanille, M. *Makromol. Chem.* **1992**, *193*, 2997-3007.
4. Klein, J.; Herzog, D.; Hajibegli, A. *Makromol. Chem., Rapid Commun.* **1985**, *6*, 675-678.
5. Matsumura, S.; Kubokawa, H.; Toshima, K. *Makromol. Chem., Rapid Commun.* **1993**, *14*, 55-58.
6. Martin, B. D.; Ampofo, S. A.; Linhart, R. J.; Dordick, J. S. *Macromolecules* **1992**, *25*, 7081-7085.
7. Whistler, R. L.; Panzer, H. P.; Roberts, H. J. *J. Org. Chem.* **1961**, *26*, 1583-1588.
8. Klein, J.; Herzog, D. *Makromol. Chem.* **1987**, *188*, 1217-1232.
9. Klein, J.; Kunz, M.; Kowalczyk, J. *Makromol. Chem.* **1990**, *191*, 517-528.
10. Flint, R. B.; Salzberg, P. L. U.S. Patent 2 016 962; *Chem. Abstr.* **1935**, *29*, 8007[8].
11. Flint, R. B.; Salzberg, P. L. U.S. Patent 2 016 963; *Chem. Abstr.* **1935**, *29*, 8007[8].
12. Yalpani, M.; Brooks, D. E. *J. Polym. Sci.: Polym. Chem. Ed.* **1985**, *23*, 1395-1405.
13. Robyt, J. F.; White, B. J. *Biochemical Techniques: Theory and Practice*; Wadsworth: California, 1987; Chapter 7.

14. Dubois, M.; Gilles, K. A.; Hamilton, J. K.; Rebers, P. A.; Smith, F. *Anal. Chem.* **1956**, *28*, 350-356.
15. An improved polymerization process in which the level of incorporation more closely matches the monomer feed has since been developed.
16. Thomaides, J. S.; Burkert, J.; Farwaha, R.; Humphreys, R. W. R.; Petersen, P. M. U.S. Patent 5 563 252.
17. Carrier, A. M.; Humphreys, R. W. R.; Petersen, P. M. U.S. Patent 5 654 198.

Chapter 17

Water Soluble Poly(acrylic acid-*co*-itaconic acid-*co*-N-vinylpyrrolidone) Materials: Optimization of Monomer Ratios for Copolymer Application in Glass-Ionomer Type Dental Restoratives

B. M. Culbertson, D. Xie, and W. M. Johnston

The Ohio State University, 305 West 12ᵗʰ Avenue, Columbus, OH 43210–1241

Water soluble, acrylic acid-itaconic acid copolymers have been utilized for some time as a material to formulate a class of dental restoratives commonly called glass-ionomers. The latter type copolymer was modified with N-vinylpyrrolidone, as a route to improve the mechanical properties of this class of dental materials. Design of experiment (DOE) techniques were utilized to determine what ratio of monomers would provide the best set of properties. It was shown that poly(acrylic acid-co-itaconic acid-co-N-vinylpyrrolidone), with monomer ratios of 7:3:1 and 8:2:1, respectively, could be utilized to formulate matrix resins for glass-ionomers, where the new matrix resins had potential for providing improved performance materials, compared to known commercial glass-ionomer systems.

The compound N-vinylpyrrolidone (NVP), as a specialty monomer, exhibits a number of unique properties. For example, NVP can be readily polymerized with free radical initiation, affording a non-ionogenic, water soluble, synthetic polymer, highly useful in the food, textiles, pharmaceutical, etc., industries. The multiplicity of uses for NVP polymers resides in the hydrophilic, non-toxicity, and high proclivity to complexation nature of the polymers. While the NVP polymer is easily soluble in water, studies have shown that water is not a thermodynamically good solvent for the homopolymer [1]. The latter is probably due to some level of compensation between the strong hydrogen bonding capability, i. e., hydrogen bonding between the cyclic amide group and solvent protons, and hydrophobic interactions that exist between water and the NVP polymer backbone and cyclic methylene groups. The strong hydrogen bonding can lead to polymer swelling, i.e., expansion of the polymeric coils, while the hydrophobic interactions could bring about compression of the coils. In any case, the NVP homopolymer is known to strongly absorb water [2], with absorption centers being the amide bond. In contrast to water soluble poly(methacrylic acid), NVP polymer does not precipitate from aqueous solution, without salts, upon heating up to 100 °C. The latter suggests the hydrophobic interactions, which would promote aggregation, do not over compensate the strong hydrophilic nature of the polymer, which results from strong hydrogen bonding. The aforesaid and other properties lead us to believe that NVP has attributes for exploitation in development of polyelectrolytes for application in glass-ionomer formulations, as further explained and shown in this paper.

The chemistry of conventional type glass-ionomers may be briefly explained as follows: Aqueous solutions of carboxylic acid functionalized copolymers are blended with basic glass powders, containing varying ratios of Si, Ca, Al and F, i. e., calcium fluoro-alumino-silicate (CaFAlSi), to obtain the formulations commonly called glass-ionomers. These mixtures harden or cure by an acid-base reaction, where the reaction occuring produces inter- and intramolecular salt-bridges, or ionic type crosslinking. In the aforesaid reaction, the salt-bridges formed are Ca or Al carboxylates, with the Al carboxylates being most desirable. The latter is a diffusion controlled process, facilitated by water present in the formulation. In the described chemistry, very few Al tricarboxylates are formed. Further, there are considerable numbers of carboxylic acid groups not tied up in salt-bridges. These acid groups are needed to maintain water solubility for the starting polymers. The early glass-ionomers employed poly(acrylic acid). But, poly(acrylic acid) was found to be unstable in aqueous solutions at elevated molecular weights, i.e., the polymer precipitated from aqueous solutions. Thus, itaconic or maleic acid was incorporated into the polymer backbone to provide disorder, enabling formulators to have stabile, aqueous solutions of the copolymers with ca. 50 % solids [3]. Hence, our concept to use NVP to modify the acrylic acid (AA)-itaconic acid (IA) copolymer was born. We theorized that a fair number of the AA or IA residues could be replaced in the copolymer backbone by NVP, maintaining water solubility for the copolymer and providing a new path to explore for generating improved performance glass-ionomers. The latter is the thrust of this study.

In our first attempt to start evaluating the concept, we examined a poly(AA-co-NVP) produced by International Specialty Products (ISP Corp., Wayne, NJ). Aqueous solutions (50 % solids) of the ISP ACP 1005 copolymer, which was reported to contain ca. 75 % AA, was blended with a glass powder used in commercial Fuji II glass-ionomer. Mixing of the glass powder (P) with the liquid (L), in the same P/L ratio (2.7:1) recommended for Fuji II, placing the mixtures in suitable molds, curing, conditioning and testing, we found the following: The AA: NVP copolymer had a compressive strength (CS) of 150 MPa, flexural strength (FS) of 20 MPa and diametral tensile strength (DTS) of 18 MPa. In contrast, the Fuji II control (test) samples, prepared under the same conditions, had CS = 183 MPa, FS = 18 MPa and DTS = 14 MPa. Both the experimental and Fuji II control specimens were conditioned in water at 37 °C for 24 h prior to testing. The preliminary results from evaluating the ISP poly(AA-co-NVP) [4] suggested to us there was an unexplored opportunity to use NVP to develop new water soluble copolymers to formulate improved performance glass-ionomers.

To shed further light on the possible opportunity to use NVP, we prepared a poly(AA-co-IA-co-NVP) material with a 7:3:1 ratio, respectively, and evaluated the FS, CS, DTS, fracture toughness (FT), working (WT) and setting (ST) times of the possible new matrix resin. Elemental analysis showed the synthesized copolymer as having 1.34 % N (theory, 1.39 % N). The IR spectra of the copolymer showed bands at 1717 and 1651 cm^{-1}, indicative, respectively, of carboxylic acid and cyclic amide groups (Table 1). The ^{1}H and ^{13}C NMR spectra, with bands at 8.32 and 12.10-12.25 and 172.0 and 176.0 ppm (Table 1), also supported the copolymer structure(s). An aqueous solution of the experimental copolymer (50 % solids) exhibited 880 cp viscosity, compared to the Fuji II liquid component (50 % solids) of 690 cp. The experimental copolymer had a gel permeation chromatography (GPC) estimated molecular weight of 44,000, with the viscosity and GPC data suggesting that the experimental and control matrix resins had comparable molecular weights. Formulating, curing, conditioning and testing of the copolymer was the same as done with copolymer AA:NVP, with the exception that all the samples were conditioned in water at 37 °C for 1 week prior to testing as previously described for poly(AA-co-NVP) [4, 5]. In this case, we determined CS to be 276 MPa and FS to be 34.0 MPa, in contrast to the Fuji II control having a CS of 205 MPa and FS of 15.0 MPa. The fracture toughness (K_{Ic}) for the experimental and Fuji II samples, determined by known ASTM methods, were 0.607 and 0.557 Mn / m$^{1/2}$, respectively. In addition, the microhardness of the cured samples, using a Leco Co. Model MVK-E M400 microhardness tester, showed the experimental product had a Knoop hardness (KHN) value of 84.3 kg/mm^2, compared to Fuji II having a KHN of 77.3 kg/mm^2 [6]. The working (WT) and setting (ST) times (min) for the experimental (WT = 5.0 and ST = 5.8) and Fuji II (WT= 2.5 and ST = 4.5) samples, showed that the NVP helped prolong the working time when preparing samples. Clearly, the experimental AA:IA:NVP (7:3:1) copolymer provided improved properties.

Having the above evidence that NVP could possibly be valuable for the design of new matrix resins for glass-ionomers, we launched the effort described in this manuscript. In this thrust, we felt the best place to start to design the new, water soluble polyelectrolyte(s) for glass-ionomers would center on the synthesis, characterization, and evaluation of poly(AA-co-IA-co-NVP), with special attention to be given to determining what was the optimum AA:IA:NVP monomers ratio and molecular weight range to use for design of an improved conventional type dental restorative.

Materials scientists often need to determine what monomer ratio in a copolymer backbone will provide optimum properties. In evaluating polymerization reactions of two or more monomers, many molar ratios may need to be prepared and studied in order to determine what is the best ratio to use, seeking to maximize some property or set of properties. We elected to start with copolymers rich in AA. Thus, having the desire to limit the number of polymers to be prepared and evaluated, we decided to use a SAS statistical program, DOE technique [7-10], to identify the optimal poly(AA-co-IA-co-NVP) copolymer(s) for glass-ionomers. The flexural strength (FS) was selected as the primary screening tool to ascertain improvements. Compressive strength (CS), along with viscosity (molecular weight optimization), would also be evaluated, to develop the glass-ionomer formulation with the best performance profile

Experimental

Materials
Monomers AA, IA and NVP, along with the potassium persulfate initiator, methyl alcohol and diethyl ether, were obtained from Aldrich Chemical Co. The AA and NVP were distilled before use.

Synthesis and Copolymer Characterization

Various ratios of AA, IA and NVP monomers were copolymerized in water with potassium persulfate initiator (2 % by wt of combined monomers) with polymerization run under nitrogen for 9-10 hr at 95 °C. The copolymers prepared (Table 2) were recovered in high yields, using standard freeze-drying techniques. Several poly(AA-co-IA) and poly(AA-co-NVP) were also prepared for the study. For purification, the copolymers were dissolved in methyl alcohol and precipitated from diethyl ether, followed by drying under vacuum. The IR spectrum of the copolymers were obtained from cast films, using a MIDAC FT IR Spectrometer. NMR (^{1}H and ^{13}C) spectra were obtained on a Bruker AM 250 MHz NMR analyzer, using deutrated methyl sulfoxide solvent and trimethylsilane (TMS) reference (Table I).

Viscosity Determinations:

Viscosities of aqueous solutions of the copolymers (Table II), with 50 % solids, were determined at ambient temperature, using a Rheometrics (RMS-800) cone and plate viscometer (Figure 1).

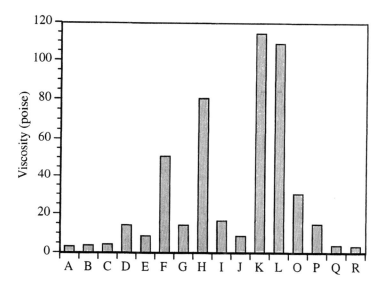

Figure 1. Viscosities of Copolymers Prepared for Study, at 50% Solids in Water.

Formulations

Glass powder used in the study were the same as that used in Fuji II glass-ionomer cement (GC America, Alsip, Illinois). The Fuji II system comes in two parts, i.e., a liquid (L) containing the copolymer and the solid glass powder (P), with recommendation for the P/L being 2.7.1. The liquid component in the experimental formulations consisted of the AA:IA:NVP, AA:IA or AA:NVP copolymers in water at 50% solids, comparable to the Fuji II liquid. The liquids were blended with the Fuji II glass powder, at the same powder / liquid (P/L) ratios recommended for the commercial Fuji II formulation. The thoroughly mixed samples were place in a mold and allowed to cure at 37 °C for 1 h at 100 % relative humidity. Upon removal from the mold, the samples were stored / conditioned in water at 37 ± 2 °C for 1 wk and then utilized for mechanical property determinations. The same procedure was used to fabricate, condition and test samples of the Fuji II control.

DOE Methodology

The methodology utilized for this study is called response-surface (RS) experimentation, with the technique consisting of four steps, defined as follows:
1) Initial data, such as FS for several formulations, are generated using a pre-planned experimental design.
2) One or more models, polynomials in this case, are fit to the data by statistical curve-fitting techniques.
3) The RS contours are examined to determine the composition regions where the optimal values of the response are predicted.
4) Additional compositions are made in the targeted region to provide additional data for refining predictions or to verify experimentally that the optimum composition has been discovered.

Models or polynomials, useful for predictions / optimizations in the DOE techniques, which may be applied to three component systems, are as follows:
1) **Linear model**: $E(y) = \beta_1 x_1 + \beta_2 x_2 + \beta_3 x_3$.
2) **Quadratic model**: $E(y) = \beta_1 x_1 + \beta_2 x_2 + \beta_3 x_3 + \beta_{12} x_1 x_2 + \beta_{13} x_1 x_3 + \beta_{23} x_2 x_3$.
3) **Special cubic model**: $E(y) = \beta_1 x_1 + \beta_2 x_2 + \beta_3 x_3 + \beta_{12} x_1 x_2 + \beta_{13} x_1 x_3 + \beta_{23} x_2 x_3 + \beta_{123} x_1 x_2 x_3$.
4) **Full cubic model**: $E(y) = \beta_1 x_1 + \beta_2 x_2 + \beta_3 x_3 + \beta_{12} x_1 x_2 + \beta_{13} x_1 x_3 + \beta_{23} x_2 x_3 + \gamma_{12} x_1 x_2 (x_1 - x_2) + \gamma_{13} x_1 x_3 (x_1 - x_3) + \gamma_{23} x_2 x_3 (x_2 - x_3) + \beta_{123} x_1 x_2 x_3$.

In the polynomials, x_1, x_2 and x_3 are molar ratios of the three components, $E(y)$ is an estimate of the property being optimized, and the βs and γs are coefficients of the model obtained by curve-fitting. Which model is used depends upon how many components there are and how many data points are available. The number of data points must be equal to or greater than the number of coefficients in any model.

Mechanical Property Testing

Samples for compressive (CS), flexural (FS) and diametral tensile strength (DTS) determinations were prepared in glass or Teflon molds. For CS the glass tubing was 4 mm diameter x 8 mm height. For DTS the sample size was 4 mm diameter x 2 mm thick. The FS and fracture toughness (FT) samples, made in a split Teflon mold, were 2 mm wide x 2 mm thick x 25 mm long. Testing was done on a Universal Testing Machine (Instron Model 4202). For the FS test, a three-point bending assembly, having a span of 20 mm between supports for FS and 16 mm between span for FT test. In all cases, tests were done at ambient conditions, following current ASTM procedures. Formulas used to calculate CS, FS and DTS were as follows: $Cs = Ppr^2$; $FS = 3PL/2bd^2$; $DTS = 2P/pDT$, where P = the load at fracture, r = radius of the sample, D = diameter of the sample, T = thickness of the sample, L = the distance between two supports, b = the breadth of the specimen, and d = the depth of the sample. For fracture toughness, $K_{1c} = [(PS)/(BW)][f(a/W)]$, the P = load at fracture, S = span between supports, B = specimen thickness, W = specimen width, a = crack length (W/2) and f = function of (a/W) whose value is obtained from the ASTM standard.

Results and Discussion

Coplymers used in this study were prepared by standard free radical polymerization in aqueous solutions. The NVP containing copolymers were prepared in high yield, using standard free radical initiation in aqueous solutions. The materials prepared were poly(AA-co-IA), poly(AA-co-NVP) and poly(AA-co-IA-co-NVP) (Tables I and II). The copolymers were easily isolated by standard freeze-drying techniques, with purification achieved by precipitation from diethyl ether, using concentrated methyl alcohol solutions of the polymers. Purity of the copolymers, i.e., absence of free monomers, was ascertained by thin layer chromatography.

TABLE I. Characterization of Selected Copolymers

Code	Copolymer	FT-IR (cm^{-1})	^1H NMR (ppm)	^{13}C NMR (ppm)
D	poly(AA-co-IA)	1726 (carboxyl on AA) 1782 (carboxyl on IA)	12.15 (carboxyl)	176.0 ($^{13}COOH$)
G	poly(AA-co-IA-co-NVP)	1720 (carboxyl on AA) 1785 (carboxyl on IA) 1649 (amide on NVP)	12.20 (carboxyl) 8.32 (amide)	176.0 ($^{13}COOH$) 172.0 ($^{13}CONR,$)
H	poly(AA-co-NVP)	1714 (carboxyl on AA) 1645 (amide on NVP)	12.20 (carboxyl) 8.30 (amide)	176.0 ($^{13}COOH$) 172.0 ($^{13}CONR,$)

The IR and NMR (^1H and ^{13}C) spectra (Table I) confirmed that the copolymer(s) contained the desired monomer residues. Elemental analysis for N % also confirmed the presence of nitrogen in the copolymer, at the levels close to that predicted for each copolymer. Typical ^1H and ^{13}C NMR chemical shifts, respectively, for the carboxyl groups were at 12.15 and 176.0 ppm, along with the chemical shifts for the amide group at 8.30 and 172.0 ppm. Poly(AA-co-IA) exhibited chemical shifts for only the carboxyl groups, whereas poly(AA-co-IA-co-NVP) and poly(AA-co-NVP) showed chemical shifts for both carboxyl and amide groups. The IR spectra for the copolymers also supported the desired structures. For the poly(AA-co-IA) copolymers there is an absorption peak at 1784 cm^{-1} for one of the itaconic acid carboxylic acid groups, in addition to a strong peak at 1726 cm^{-1} for the acrylic acid residue. For poly(AA-co-NVP) copolymers there exists a broad peak at 1647 cm^{-1} for an amide group in the pyrrolidone ring, along with one at 1726 cm^{-1} for the acrylic acid. For poly(AA-co-IA-co-NVP) terpolymers there exists all three, acrylic and itaconic acids and NVP amide peaks.

Looking at the various AA, IA and NVP combinations of monomers, only seven combinations exist, i.e., three homopolymers, three copolymers and one terpolymer. But, only four polymers would be considered useful for glass-ionomers, namely poly(AA), poly(AA-co-IA), poly(AA-co-NVP), and poly(AA-co-IA-co-NVP). The poly(IA), poly(NVP) and poly(IA-co-NVP) would not be considered useful. The poly(AA) is generally used at MW <40,000, due to polyacid precipitation from aqueous solution on standing. Itaconic acid (IA) tends to copolymerize with other monomers but homopolymerize with difficulty. The poly(NVP), for our study would be considered completely useless, due to lack of any carboxylic acid groups. Poly(AA-co-IA) is currently being used in several commercial glass-ionomers, with Fuji II being an example . Earlier [4-6], we had shown that poly(AA-co-NVP) and poly(AA-co-IA-co-NVP) provided some reinforcement of the concept that NVP modified acrylic acid copolymers could offer a path to glass-ionomer restoratives with improved properties. In summary, all our preliminary studies [4-6] suggested that NVP could be very attractive to study in modification of matrix resins for glass-ionomers, with focus needed to determine the optimum monomer ratio in the copolymers backbone.

Statistical design of experiment (DOE) is an efficient procedure for finding the optimum molar ratio for copolymers having the best property profile. Based on the concepts of response-surface (RS) methodology, developed by Box and Wilson [11], there are four models or polynominals (Table III) useful in our study. For three components, in general, if there are seven to nine experimental data points, the linear, quadratic and special cubic will be applicable for use in predictions. If there are ten or more data points, the full cubic model will also be applicable. At the start of the effort, one prepares a fair number of copolymers with different AA:IA:NVP ratios and tests for a property one wishes to optimize, with the data fit to the statistical models. Based on the models, new copolymers, with different ratios, are prepared and tested for the desired property improvement. This type procedure significantly lowers the number of copolymers that needs to be prepared and evaluated, in order to identify the ratio needed to give the best mechanical property.

TABLE II. Molar Ratios and Average Flexural Strength Values

| Code | Molar Ratio | | | Flexural Strength (MPa) (S.D.) | CS (MPa) (S.D.) | DTS (MPa) (S.D.) | Modeling Step |
	AA	IA	NVP				
A	1	1	1	12.68 (2.048)	69.39±3.811	6.331±0.661	Step 1
B	5.5	4.5	1	13.76 (1.899)	133.7±10.83	13.28±1.679	Step 1
C	3	7	1	14.58 (2.326)	142.8±9.490	10.67±0.878	Step 1
D	7	3	0	16.85 (1.108)	177.7±10.48	18.19±1.511	Step 1
E	7	3	1	19.47 (1.147)	195.8±7.941	16.83±1.975	Step 1
F	8.5	1.5	1	26.60 (3.312)	163.8±9.413	21.90±1.414	Step 1
G	7	1	3	31.40 (2.705)	143.4±7.875	15.54±0.944	Step 1
H	7	0	3	35.75 (4.118)	134.5±6.748	17.29±0.936	Step 1
I	6	2.7	2	29.86 (6.035)	136.6±11.10	13.26±1.583	Step 2
J	8	1	2.5	16.79 (2.909)	107.9±9.230	11.33±0.950	Step 2
K	5	0	2	21.25 (3.367)	121.2±11.62	17.04±0.837	Step 2
L	3	0	2	37.86 (5.469)	112.1±6.819	14.34±0.875	Step 2
M	6	0	1	-	-	-	Step 2
N	16	0	1	-	-	-	Step 2
O	6	1	4.5	15.82 (2.822)	88.67±5.829	9.051±0.373	Step 3
P	6	1	3.4	17.92 (3.945)	100.0±8.875	9.380±0.732	Step 3
Q	1.4	2.6	1	10.16 (1.383)	70.25±6.377	6.522±0.321	Step 3
R	2.2	1.9	1	9.301 (2.024)	81.97±7.307	6.602±0.572	Step 3

Flexural strength (FS) was selected as the primary mechanical property we wished to optimize in the experimental glass-ionomers, since FS is one of the properties needing significant improvement. In fact, Prosser et al. [12] suggested that the most appropriate measure of the strength of glass-ionomer cements was FS, since a material could fail by separation of the planes of atoms (i.e., tensile failure), or by slipping of the planes of atoms (i.e., shear failure). The FS test is a collective measurement of three types of stresses, simultaneously, i.e., tensile at one surface of the specimen beam, compression at the other surface, and shear in the direction which is parallel to the load stress [12]. The work of Williams et al. [13] also suggested, from a clinical perspective, the FS test would be most appropriate to optimize. Hence, we choose FS as the primary property to optimize in establishing the best AA:IA:NVP ratio to function as a matrix resin.

For our initial experiments, i. e., Step 1,we chose eight different ratios in the first set of experiments, as shown in Table II (A-H). The eight different ratios selected were based on previous observations [4,5] as well as studies by others [14, 15], and the knowledge that the copolymers must have a high level of carboxylic acid groups. After synthesis of the copolymers, their purification, fabrication into test specimens and conditioning, the FS for the eight copolymers were obtained (Table II, Figure 2). Only the linear, quadratic and special models were used, since only eight sets of data points were collected. The plotted RS contours for the special cubic model suggested we should prepare ratios more rich in AA and NVP, as shown in Table III, with the ratio to study for AA:IA:NVP being 0.66:0.007:0.33, which could possibly provide a material having a FS of 37.3 MPa. The RS contours for the quadratic model predicted a AA:IA:NVP ratio of 0.773:0.007:0.22, which could possibly provide a FS of 37.5 MPa (Table III). The linear model RS contours indicated the same ratio as the special cubic model, with a predicted FS of 31.6 MPa. Based on the latter models, six additional copolymers were prepared (I - N), purified and fabricated, where possible, into test specimens, followed by conditioning, and evaluating for FS (Table II). As shown, these additional six experiments (Table II), followed by detection of their FS (Figure 2), gave us a new set of strength data.

TABLE III. Predicted Molar Ratios for Each Modeling Step

Model	Predicted Point			Expected	
	AA	IA	NVP	FS (MPa)	Step
Linear	0.663	0.007	0.330	31.60	1
Quadratic	0.773	0.007	0.220	37.50	1
Special Cubic	0.663	0.007	0.330	37.30	1
Full Cubic	-	-	-	-	1
Linear	0.603	0.007	0.390	30.70	2
Quadratic	0.603	0.007	0.390	34.60	2
Special Cubic	0.523	0.087	0.390	42.70	2
Full Cubic	0.273	0.527	0.200	73.60	2
Linear	0.773	0.007	0.220	28.00	3
Quadratic	0.733	0.007	0.260	29.70	3
Special Cubic	0.723	0.007	0.270	29.80	3
Full Cubic	0.603	0.007	0.390	31.50	3

The RS contour diagram for the modeling in Step 2 (Table II), i. e., using a second set of six molar ratios yielded two copolymers, M and N, which could not be evaluated, since their aqueous solutions were not stabile, i.e., water solutions of the copolymers were hydrogels. It was observed that a very high amount of AA in the copolymer would cause formation of unstable aqueous solutions. The latter is akin to what is reported for high MW acrylic acid polymers [3, 15]. As shown in Table III, the special cubic model contour suggested the copolymer should have more of both the NVP and AA, i.e., selecting a ratio of 0.523:0.087:0.39 would provide a FS of 42.7 MPa. The full cubic model contour pointed toward compositions having more IA, less NVP and AA, i. e., 0.273:0.527:0.2, with predicted FS of 73.6 MPa. At the same time, the linear and quadratic contour surfaces indicated similar trends, with both in the direction of a 0.603:0.007:0.39 ratio, with an expected FS strengths of 30.7 and 34.6 MPa, respectively. The four data points at O (0.523:0.087:0.39) and P (0.690:0.088:0.22), based on the special cubic, linear, and quadratic models, and at Q (0.273:0.527:0.20) and R (0.430:0.370:0.20) , based on the full cubic model.

In Step 3, the detected FS values did not show any improvements, over the results predicted (Table II) and shown in Table II and Figure 2. The new directions predicted by the four models all pointed towards more AA or NVP, with projected FS values of 28.0 - 31.5 MPa, as shown in Table 3. Since there were no further significant improvement in FS when the molar ratios were moved to more NVP or AA, the optimization process, using FS as a primary screening tool, was stopped. At this point, further examination of the copolymers were undertaken, looking at handeling, CS, DTS, etc., properties for the formulations. In total, we prepared and examined eighteen different copolymers, with all copolymers having molar ratios concentrated mostly toward the AA corner.

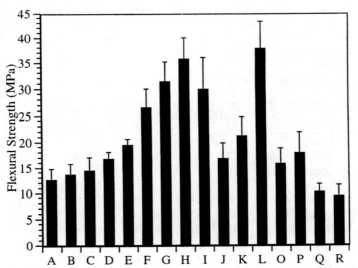

Figure 2. Flexural Strength (FS) for Glass-Ionomer (Copolymer) Formulations in Study, With Error Bars and Standard Deviations Obtained.

In preparing and working with glass-ionomer formulations it is very important to consider both working and setting properties. In particular, it is important to not have too high a working viscosity, allowing one to readily blend the aqueous solution of the copolymer (50 % solids) with the glass powder, at P/L ratios in the vicinity of 2.7/1. It is important to note here, addition of NVP to the poly(AA-co-IA) backbone allows for easier blending of the glass powder, with smoother mixtures obtained. Figure 1 shows the viscosities (50 % solids) of the various copolymers prepared in this study. Copolymers L and H had the highest FS (37.9 and 35.7 MPa), along with the highest viscosities of 108.8 and 80.2 poise, which could possible be viscocities too high for clinical application. Copolymers A, B, C, J, Q and R showed lower FS, which might be attributed, in some degree, to their lower viscosities or lower molecular weights. Copolymers D, P and O had fairly reasonable working viscosities, but still showed somewhat lower FS values. Copolymer K showed the highest viscosity (114.3 poise), with an FS value of 21.25 MPa. Among the other copolymers (E, F, G and D), copolymer G (7:1:3) had the highest FS value at 31.4 MPa, and lower viscosity of 14.3 poise, while the others showed FS values of 19.47-29.86 MPa, along with reasonable viscosities of 14.27-50.90 poise. Based on the FS determinations and the very limited viscosity studies, we concluded that the molar ratio of 7:1:3 (AA:IA:NVP) was near optimum for the copolymer.

While the FS is a very important physical property for the brittle glass-ionomer systems, there are several other important properties that would also need optimization, such as CS and DTS. With this in mind, and seeking to see if the optimum FS ratio of 7:1:3 would also be optimum for CS and DTS, we determined the CS and DTS values for all the glass-ionomers produced with our copolymers, with results shown in Table II. As shown, the copolymer with a 7:1:3 ratio did not show the highest CS and DTS. The highest DTS was found for copolymer F (8.5:1.5:1), while copolymer E (7:3:1) exhibited the highes CS value. From these results, we postulate that simultaneous optimization of FS, CS and DTS values for the poly(AA-co-IA-co-NVP) matrix resin would be somewhere between 7:3:1 and 8.5:1.5:1, or at ca. 8:2:1.

The data collected tends to show that more NVP leads to higher FS and less NVP brings about higher CS and DTS values. While additional studies are needed to understand this effect, it is possibly related to less opportunity for salt-bridge formation to occur with higher NVP content, reducing CS and DTS values. Considering the CS test, more brittle materials tend to have higher CS, since compression is easy to close the microcrack flaws or defects in the material [16]. This may be why higher CS is found at lower NVP, i.e., less ionic crosslinking is produced on curing, making for a less brittle composite. In the systems containing more NVP, the NVP residues may possibly form some strong hydrophilic domains, which could inhibit the planes of atoms to slip, slide or separate in some fashion. In any case, additional studies need to be done to understand the mechanism for NVP enhancement of the FS and reduction of CS and DTS values.

Conclusions

As part of a program to prepare new matrix resins for the formulation of improved dental restoratives, called glass-ionomers, we have explored the use of N-vinylpyrrolidone (NVP) as a modifier for poly(acrylic acid-co-itaconic acid) materials currently used. The optimal molar ratio of the copolymers composed of the three monomers acrylic acid, itaconic acid and NVP was explored, using statistcal design, or design of experiment (DOE) techniques. It was found that the poly(acrylic acid-co-itaconic acid-co-N-vinylpyrrolidone) having a 7:1:3 monomers ratio, respectively, exhibited the best flexural strength (FS). However, the 7:1:3 ratio was not the best for the compressive (CS) and diametral tensile (DTS) strengths. As a compromise, trying to maximize CS and DTS as well as FS, the molar ratio of 8:2:1 (AA:IA:NVP) was considered optimum. It was shown that the statistical design of experiments (DOE) techniques were very useful for identification of the best molar ratios and point out the synthesis direction for a quicker determination of the optimum ratios for properties such FS, CS and DTS.

In future studies we need to focus on optimization of molecular weights for the 7:1:3 or 8:2:1 copolymers, seeking to determine the highest molecular weight we can tolerate and still obtain good mixing with the glass powders. Also, we need to look at how the copolymers are prepared, to ensure the most statistical distribution of monomers in the polymer backbone is achieved. The aforesaid factors will also influence how well the copolymers perform for improving the properties of conventional glass-ionomers.

Acknowledgement

The support of NIH (grant DE 11682) and The Ohio State University College of Dentistry, to do this study, was greatly appreciated.

Literature Cited

1. Levy, G. B. ; Frank, H. P. J. Polym. Sci. 1955, 17, p. 247.
2. Dole, M. ; Faller, I. L. J. Am. Chem. Soc. 1950, 72, p. 414.
3. Wilson, A. D.; McLean, J. W. Glass-Ionomer Cement, Quintessence Publishing Co., Chicago, IL, 1988 Chapters 2 and 3, pp. 21-54.
4. Xie, D. ; Culbertson, B. M. J. Dent Res. 1996, 75, p. 298.
5. Xie, D. ; Culbertson, B. M. J. Dent Res. 1997, 76, p. 74.
6. Xie, D. ; Culbertson, B. C.; Wang, G., J.M.S.-Pure Appl. Chem. 1998, A35(4), pp. 547-561.
7. Hahn, G. J.; Morgan, C. B. Chemtech, 1998, pp. 664-669.
8. Cawse, J. N.; Izadi, N. Today's Chemist at Work, 1992, pp. 24-28.
9. Snee, R. D. Chemtech, 1979, pp. 702-710.
10. Hunter, J. S. Chemtech, 1987, pp. 167-169.
11. Box, G. E. P.; Wilson, K. B., J. of 6the Royal Statistical Society, 1951, Series B, 13, pp. 1-45.
12. Prosser, H. J. ; Powis, D. R.; Wilson, A. D. J. Dent Res. 1986, 65(20), pp. 146-148.
13. Williams, J. A.; Billington, R. W. ; Pearson, G. J. Br. Dent J. 1992, 172, pp. 279-282.
14. Kent, B. E.; Lewis, B. G.; Wilson, A. D. J. Dent Res. 1979, 58(6), p.1619.
15. Crisp, S.; Kent, B. E.; Lewis, B.G.; Ferner, A. J.; Wilson, A. D. J. Dent Res. 1980, 59(6), pp. 1055-1063.
16. Shackelford, J. F., in Introduction to Materials Science for Engineers, Macmillian Publ. Co., 1988.

Author Index

Subject Index